ADVENTURES OF
A ZOOLOGIST

ALSO BY VICTOR B. SCHEFFER

The Year of the Whale
The Little Calf
The Year of the Seal
The Seeing Eye
A Voice for Wildlife
A Natural History of Marine Mammals

Adventures of a Zoologist

~~~ VICTOR B. SCHEFFER ~~~

CHARLES SCRIBNER'S SONS / NEW YORK

QL
31
.S28
A34
1980

Copyright © 1980 Victor B. Scheffer

Library of Congress Cataloging in Publication Data

Scheffer, Victor B.
Adventures of a zoologist.

Includes index.
1. Scheffer, Victor B.   2. Zoologists—Washington (State)—Biography.   I. Title.
QL31.S28A34   599'.0092'4 [B]   79-27463
ISBN 0-684-16439-6

This book published simultaneously in the United States of America and in Canada—
Copyright under the Berne Convention

All rights reserved. No part of this book may be reproduced in any form without the
permission of Charles Scribner's Sons.

1 3 5 7 9 11 13 15 17 19   H/C   20 18 16 14 12 10 8 6 4 2

Printed in the United States of America

> Either love . . . is something which we share with animals or it is something which does not really exist within us.
> —Joseph Wood Krutch,
> *The Voice of the Desert*

CONTENTS

INTRODUCTION　*xi*

1. APPRENTICE WORK　*1*
   Searching for pests. The life zones of Rainier. Aboard the *Catalyst*

2. THE ALEUTIAN EXPEDITION　*14*
   Aboard the *Brown Bear*. Birds, foxes, and sea otters. Last of the whaling stations. Sorting out the collections. Was it all worthwhile? The Aleutian Islands then and now. Olaus Murie

3. STUDIES OF LAND ANIMALS　*34*
   The mysterious Mima mounds. The mountain beaver, a living fossil. Tracking a lost herd of deer. Last of the sea-otter hunters. A gray wolf skin. An experience among cannibals

4. TO THE FUR-SEAL ISLANDS　*47*
   First trip to the Pribilofs. The plight of the Aleuts. Overwhelmed by seals. A roundup and a branding. Fresh eggs by the thousands. Interludes of beauty. The rise and fall of a reindeer herd. Threatened collapse of the fur-seal treaty

5. LOCKED ONTO SEA MAMMALS　*66*
   Scientists in the city of Washington. Training to specialize. Planning the *Black Douglas* expeditions. Research in wartime. Scientists in California

6 BACK TO THE SEAL ISLANDS *84*

   The health of the seals. How old is a seal? Visitors. The *Black Douglas* restored. A breakthrough in counting seals. Outstaying a harem master. New knowledge of sea otters

7. A NEW FUR-SEAL TREATY *107*

   The treaty is signed. Aiming for maximum yield. Visit to a Russian seal island. Final years in the Seattle laboratory. Virility of the bull seals

8. TEACHING *120*

   Organizing, defining, and compiling. Teaching at the University of Washington. At the College of the Cayman Islands

9. COUNSELING *132*

   Efforts in wildland preservation. Hiking with Justice Douglas. The Nature Conservancy. A secret mission. Among walruses in the Bering Sea. Among manatees in Guyana

10. SUPERVISING *159*

    Dreaming of a sea mammal center. The early conferences on sea mammals. Chairman of the Marine Mammal Commission

11. WRITING *171*

    Seals, Sea Lions, and Walruses. The Year of the Whale. The Year of the Seal. Dabbling in art. *A Voice for Wildlife. A Natural History of Marine Mammals*

12. A MORAL ENDING *184*

    Credo. Zoology and morality

    INDEX *195*

INTRODUCTION

THIS book is part autobiography and part history. It tells of my search for a career in zoology; then of pursuing that career and, in the end, recalls its livelier moments.

I conclude that zoology can be rewarding—less perhaps in money than in the special values enjoyed by poets, painters, scholars, and explorers. I shall be pleased if some readers who are on the point of choosing a career will be persuaded that the life of a zoologist can be exciting as well as difficult, humane as well as adventurous.

The word *zoology* means broadly the study of animals, although it usually leads to a specialty which could be embryology (the study of early growth), genetics (the study of inheritance), pathology (the study of disease), or any of the scores of specialties concerned with fundamentals common to all animal life. Or the specialty could deal with a specific group of animals. It could be entomology (the study of insects), ornithology (the study of birds), mammalogy (the study of mammals), or other.

I am simultaneously a mammalogist, a marine mammalogist, and a wildlife management zoologist. I trust that these terms will become clearer as the narrative unfolds.

The kind of zoology I shall tell about is not the kind in which DNA molecules are unraveled, or fetal rats are brought to term in bottles, or newts are cloned in the image of "father." Rather, it is the outdoor kind—the wet and dirty kind—in which men and women brace their feet on rolling decks, or search on sandy beaches for stranded marine animals, or hide among rocks to spy on seals, or spend long hours at some aquarium pool clocking the behavior of a captive animal.

INTRODUCTION

As an informal history, the book illuminates the ongoing campaign to conserve the beasts of the sea—the seals, sea otters, whales, dolphins, porpoises, manatees, and dugongs. These sleek, flippered animals include more than a hundred species and perhaps a hundred million individuals. In the role of zoologist I have studied the lives of certain ones and in the role of civil servant I have worked within agencies responsible for protecting the wild populations (or stocks) of others.

I write of the fifty years from 1928 to 1977, a half-century marked by a worldwide surge of interest in sea mammals (and indeed in all the wildlife of our planet). The latest international directory of specialists who are studying sea mammals contains 607 names. In 1928, had there been such a list, it would have contained no more than a few dozen.

The events in the following pages are narrated only roughly in chronological order. A historic trend is often followed from start to finish regardless of its time, as a thread might be pulled from a fabric and traced to its end. In the year 1928, when the story opens, I first earned pay as a zoologist. And the whaling industry—the richest of all the sea mammal industries—was expanding rapidly. That year the hunters killed 9,627 blue whales, nearly as many as all those now alive (about 13,000). The industry was just beginning to search for scientific knowledge of the whales and for ways of checking its own bloody spree.

The American Society of Mammalogists had appointed in 1921 a standing committee charged with the responsibility, among others, of recommending measures to protect whales. Its first members were Barton W. Evermann (California Academy of Sciences), Gerrit S. Miller, Jr. (U.S. National Museum), Robert Cushman Murphy (American Museum of Natural History), and Theodore S. Palmer (U.S. Biological Survey). In later years, many North American zoologists concerned with sea mammal protection, myself among them, served on that committee. In 1930 international whale experts met under the sponsorship of the League of Nations to lay the groundwork for the first whale conservation treaty. One of the delegates was Remington Kellogg of the National Museum, a man who was later to counsel me wisely on many occasions.

On those rare occasions when I think back to the climate of

INTRODUCTION

opinion that surrounded zoologists in the 1930s I realize that few of us foresaw the emergence of two public concerns which now loom in our vision and which govern our actions. These are respect for nature and naturalness (the environmental-ecological idea) and respect for life itself (the anti-kill idea). In the 1930s we supposed that we could "manage" nature and "improve" or "reclaim" the earth's ancient, time-tested organic systems. Now we are less sure. We spoke of living animals as "resources." In the interest of science, and with a clear conscience, we hot-iron branded them, trapped them in painful devices, and shot, clubbed, or harpooned them.

I write these words not in self-reproach but simply to explain an older zoology. They offer a starting point from which the changes of the next half-century will be measured.

~ As I spin out the history of sea mammal conservation I shall try also to demonstrate that zoology in the broad sense has become increasingly concerned with its relevance to human needs. Its new focus is a consequence of increasing human populations and shrinking per-capita budgets for pure or arcane (some would say impractical) research. Most importantly, zoologists are beginning to see their own roles with respect to the fragile ecosystems of Earth. They are gaining a new insight, which is bringing zoologists and ecologists together. And I would like to think that zoologists by and large are becoming a little more human and humane—indeed more humorous, if you will concede that a sense of humor can be something close to a sense of values.

The source materials for this book include personal publications, about three-quarters of which are technical and one-quarter nontechnical, scattered through books and journals from *American Fabrics* to *Zeitschrift für Säugetierkunde.* The sources include, too, my catalog of photographs dating from 1938, and notes that my mother (Celia Esther Scheffer, 1878–1959) copied during her final years from her lifelong diaries. Then she burned the diaries—"for fear someone will read them." A curious act. They could not have revealed sins committed, for there were none. They doubtless reflected days of despair, but which one of us has not suffered those?

ADVENTURES OF
A ZOOLOGIST

1

Apprentice Work

FORTY miles from the University of Washington in Seattle there lies a town bearing an Indian name which people often mispronounce because they are smiling. Spelled Puyallup, the name is spoken *pew ál up.* From that little town I entered the University in 1925, and at the University I took a course in entomology under an exceptional teacher, Trevor Kincaid (1872–1970). He was a zoologist. I can see him now in a sunlit room of my memory, standing in Science Hall with chalk dust on his pants, pawing the air to show us how the male water flea transfers sperm to the female with his tiny fifth appendages.

During the eleven-year period 1925–36, when I was either his student or his teaching assistant, he subtly convinced me that men and women can profit more by *wondering* about things and *marveling* at them than by striving to *own* them. He made zoology seem more important than other careers that I could have adopted in the hope of becoming rich. He nourished in my youthful mind a feeling of community with all living things, a feeling of common-kindness. Had I never felt that, I should later have found zoology a dull occupation.

Plain *zoology* as a word standing alone is one seldom seen in print nowadays, partly because zoology has splintered into many disciplines, each with its own name (all, of course, devoted to understanding animals). The behavioral zoologist or ethologist, for example, may study how a honeybee communicates by sign language to others in the hive that it has found a source of food. The animal physiologist may study how a ground squirrel, hibernating in April beneath an alpine snowfield, knows that spring has come and that it can now tunnel up to daylight and find the first new blades of grass. The systematic zoologist devotes his career to studying the geographic distribution of, and variation in, animals. (I shall dwell later on his work, which aims to classify the more than one million kinds of animals and to label them in a universal language.) The work of the systematist is closely related to that of the animal evolutionist and animal paleontologist, who search for evidence of the ancient origins of modern species. Still another kind of zoologist, the population dynamicist, studies the reasons behind population changes such as the explosions of lemmings, muskrats, field mice, and Arctic hares.

There is a second reason why the word *zoology* is seldom used alone. It is increasingly recognized as merely part of two larger fields—biology (the study of all life) and ecology (the study of life in the context of its environment). Zoology is becoming harder and harder to isolate. College textbooks which, fifty years ago, would have been published under the title "Zoology" are now appearing as "Biology of Animals," or "Life: An Introduction to Biology," or "Life: The Individual, the Species," or similar titles.

And, broadly speaking, there is a kind of zoology known as "practical" (or applied, or mission-oriented) in distinction to that known as "pure." The boundary between them is not sharp.

The practical reasons for studying animals must surely number in the tens of thousands. They range from interest in developing new medical drugs through study of laboratory mice to interest in saving bowhead whales from extinction. My old professor, Kincaid, was sent by the United States Government in the early 1900s to Japan to collect live parasites of the gypsy moth, a devastating forest pest. His mission was one of the earliest efforts to control a pest by nonchemical means, that is, to control life with life. Later, he revisited Japan and brought back larvae of the giant oyster *(Ostrea gigas)*, thereby starting a new shellfishery in the United States. Thanks to him, I now enjoy gathering oysters on the beach below my Puget Sound home.

1. Prof. Trevor Kincaid on University of Washington zoology field trip, Green River Gorge, Washington, early 1930s.

ADVENTURES OF A ZOOLOGIST

The pure reasons for studying animals are simply to find Truth for the sake of Truth. When zoologists discovered in the 1960s that growth layers in fossil corals and seashells preserve faithfully the record of a distant era when Earth's rotation was slower than now, they contributed nothing at all to make human life easier, but they did make it richer. Pure zoology resembles pure art in that it aims mainly to deepen one's sense of wonder. This is perhaps enough, for it is imagination that makes us human.

Searching for Pests

But I started to tell about the entomology class. It led in the summer of 1928 to my first appointment in zoology, as a temporary field assistant at the United States Bureau of Entomology's field station in Puyallup. Two other young men and I were hired to slice daffodil bulbs and to count in each the number of bulb-fly grubs *(Lampetia)* that had survived experimental treatment. Some bulbs had been soaked in hot water, while others had been exposed to the fumes of hydrogen cyanide. After World War I, bulb growing, based on Holland stock, had become highly profitable. Unfortunately the first imports carried the bulb fly but not the parasites which, in Holland, had kept the fly under control.

In the 1950s control of the fly shifted to methods that entailed soaking the bulbs in, or drenching the soil with, one of the new wonder chemicals such as aldrin, chlordane, or heptachlor. Still later it was found that insecticides in this family of chemicals are ineffective for sustained use, as well as highly dangerous to nontarget species. The bulb growers turned to other methods of control and also learned to live with bulb flies in small numbers. Even in my day, the flies infested fewer than one bulb in twenty.

As Ian Mackay, British newsman, has written, "This sorry world is big enough to contain a fly or two."

APPRENTICE WORK

The Life Zones of Rainier

In the summer of 1930 my internship in zoology took a pleasant turn. A temporary job as ranger-naturalist at Mount Rainier National Park was offered. I took it, and continued to work at Rainier through the five summers of 1930–34 under the direction of C. Frank Brockman, chief naturalist. Our work was called "interpretation of natural values" for the benefit of visitors. Nowadays, the Park Service educational complex of museums, signposted trails, and evening lectures is familiar to most Americans, whereas fifty years ago it was still a novelty.

Delightful it was up there in the mountain meadows, breathing the sharp fragrance of yellow cedars and alpine firs as one guided visitors along the nature trails. Small children ran about, stuffing the jaws of chipmunks and jays with chocolates. Black bears tipped over garbage cans and, searching for goodies, fell through the canvas tops of automobiles. They broke into the horse stable and ate oats; they stole pies from picnic tables. One day a monkey which had escaped from a visitor's car walked into the rangers' toilet and scared the hell out of a naturalist who was occupied there.

In the 1930s the problems of parkland management, today grown so enormous and so vexing, were just being recognized. The National Park Service was still unsure what the public wanted (or would tolerate) and how much abuse the land and its vegetative cover could withstand. Government policy was still framed in the simplistic style of a needlepoint sampler:

THE PARKS ARE CLIMAX EXAMPLES OF NATURAL WONDERS
TO BE PRESERVED UNSPOILED FOR ALL TIME

Ah yes.

We found the body of a bear, poisoned by a hotel employee who had reached the limit of his patience. We tried to thin out the "beggar bear" population by live trapping a few bears and hauling them miles away. All came back. We found chipmunks lying on the ground, dying of sylvatic plague (a mild form of bubonic), and we wondered whether candy and other unnatural foods had predisposed them to disease. A small girl was gored by a buck deer. Although he was simply acting out his territorial imperative, we shot him to appease the parents of the child.

One of our group was Alton A. Lindsey, who left us in 1933 to serve as a biologist with Admiral Byrd on the Second Antarctic Expedition. He later went on to a distinguished career in botany at Purdue University. In a recent letter to me he reflects that our public lectures back in the thirties overemphasized the importance of *naming* plants and animals. Nowadays, he believes, there is an "overall principles, ecosystem approach in the best interpretative works." In short, the why, as well as the what, of nature.

He's right, I suppose, yet Henry van Dyke, American essayist of my father's day, thought otherwise. "To be able to call the plants by name," he wrote, in 1899 in *Little Rivers: A Book of Essays in Profitable Idleness*, "makes them a hundred-fold more sweet and intimate. Naming things is one of the oldest and simplest of human pastimes."

Frank Brockman, who supervised all the ranger-naturalists on Rainier, later resigned to teach forestry at the University of Washington. When I first knew him more than forty years ago I supposed that he was discontented in his career, for he fumed and sputtered at the imperfections of the Park Service. Later I realized that he was a purist. So firmly did he believe in high goals for the Service that he simply couldn't wait for them to be reached via normal bureaucratic channels. He was not, in fact, discontented but was *enjoying* his role as an activist in the continuing struggle for better park management. Though he seemed cranky, he had a vision, and he made people move.

Fortunately for the Park Service, another man with vision had become its assistant director on July 1, 1930, and had created the branch of education and research. He was Harold C. Bryant, fresh from the Museum of Vertebrate Zoology at Berkeley, California. Through his efforts, the position of chief naturalist in all the national parks was raised to the same level as that of chief ranger, just below that of park superintendent. Law enforcement, fire fighting, trail building, search and rescue, and

APPRENTICE WORK

similar functions long regarded as of the highest importance were now classified no higher than those of displaying and interpreting the natural values of the parks.

~ My summers at Rainier brought two rewards that I was not fully to appreciate until many years later.

First, I gained practice in explaining scientific facts about geology, botany, and zoology. It led me into science writing, an avocation I still enjoy but one some zoologists will claim is a waste of time that otherwise might have been spent in the pursuit of useful knowledge. Some will even claim that science writing, insofar as it exploits the work of others, is parasitic. I shall have more to say about this later.

The second reward was clearer insight into the wholeness and complexity of natural ecosystems. On the windward side of Rainier fifty feet of snow falls annually. Its runoff cools the feet of mountain goats and delights the bobbing water ouzels. It soaks into fields where avalanche lilies, lupines, paintbrushes, and pasqueflowers blaze in the short summer days. Along runways through the flowers, field mice, pikas, and marmots guard themselves by day against hawks and golden eagles, and by night against foxes and coyotes.

The slopes of Rainier, from its evergreen base to its shining crown 14,410 feet above sea level, are a great posterboard on which are displayed four life zones. (A life zone is a geographic belt having characteristic plants and animals and limited by temperature.) The zones are stacked like layers of a cake. As one climbs through them one sees faunas and floras much like those that one would see in traveling south-to-north from Puget Sound to Alaska. Although I had read about life zones, I did not comprehend them until I had hiked from a Douglas-fir lowland, through a Canadian forest of silver fir, across an alpine meadow, and into the rocky barrens above timberline—the zone of goat and ptarmigan.

C. Hart Merriam, first chief of the United States Biological Survey, was also the first person to arouse widespread interest in the life-zone concept. Stationed in the city of Washington, he had been receiving, for a decade, reports from his field men who were systematically collecting plants and animals throughout the United States. (I imagine that Merriam personally saw most of the herbarium sheets and the animal skins and skulls that passed through his office en route to the National Museum.) He came to realize that certain species of plants and animals form associations (floras and faunas) and that these can usefully be classified into broad, transcontinental zones. His 1898 publication, "Life Zones and Crop Zones of the United States," was one of the first introductions to the landtypes which today are termed the world's eight biomes. Their names are self-descriptive: the tundra, northern coniferous forest, temperate deciduous forest, tropical forest, grassland, desert, freshwater, and saltwater (or marine) biomes. A biome is largely shaped by its climate, while its plants are a clue to that climate and its animals are a clue to the kinds of food and shelter provided by the plants.

The life-zone classification introduced by Merriam was useful in its day as an effort to explain those associations of living things which any traveler with open eyes can see, particularly among the mountains of the American West. It failed, however, to mirror the full complexity of natural ecosystems. It took into account imperfectly the many factors—including rainfall, snowfall, frost, wind, sunshine, soil type, altitude, slope, associated species of organisms, and the passage of time itself—which arrange the spatial patterns of life. As a consequence, modern books dealing with ecology tend to dismiss the life-zone concept with brief notice.

I never met Merriam, although once in 1941 I enjoyed Sunday dinner with his son-in-law, Vernon Bailey, in the Merriam home. On that particular day, the old gentleman himself —then eighty-six—was out. Both Merriam and Bailey died in 1942.

I like to think of Merriam not as among the last of the naturalists but as among the first of the environmentalists. He surely understood that every environment on earth is a language that is continually telling of the past and the present and (if one listens carefully) is predicting the future.

APPRENTICE WORK

Aboard the Catalyst

My summers at Rainier were interrupted briefly in 1932 by a trip to Alaska. Robert C. Miller, then professor of zoology at the University of Washington, offered me an unpaid job as "zoological deckhand" on the research vessel *Catalyst*. The *Catalyst* was a seventy-foot, top-heavy vessel newly commissioned by the University's department of oceanography. Miller invited me to help him take plankton hauls and bottom samples. I jumped at the chance, partly because the ship was headed for Alaska, where I had never been, and partly because I liked and respected Miller.

We were aboard ship for about a month in late summer. Along the beautiful Inside Passage, which ends at Skagway, we saw virgin forests of spruce and cedar rising to the sky, their green flanks dripping with waterfalls. As our ship turned into one quiet channel after another, waterfowl would rise in alarm, leaving traces on the mirror sea. Dall porpoises, bold in black and white, would often appear from nowhere and play for a while alongside the ship's bow. Each would say "CHUFF!" as it blew at the surface before diving an instant later. One magical evening we watched porpoises writing their brilliant green signatures underwater as they streaked through swarms of luminescent plankton.

The *Catalyst* trip was an example of first-stage, or exploratory, zoology, the main objective of which is to learn what species are present. In later stages, population numbers are estimated and a myriad of life-history facts, such as breeding biology and feeding habits, are obtained. Bob Miller and his colleague, John E. Guberlet, collected samples of plankton and preserved them in formalin for later identification. They also operated a mechanical claw that grabbed mud from the sea bottom and dumped it, teeming with crustaceans, mollusks, worms, hydroids, and starfishes, on the *Catalyst*'s deck. One special kind of gray, buttery, glacial mud, rich in life but awfully hard to remove from one's clothing, we labeled "Gooby Mud" in Guberlet's honor.

When we were not sampling the sea we were looking at birds—gulls, albatrosses, shearwaters, petrels, and other sea-fowl. The black-footed albatross, or goony, is symbolic of the open North Pacific. Gliding silently on wings that span seven feet, it rides on wind currents deflected upward by the waves. Often, though, I had to appreciate these birds while suppressing thoughts of seasickness. And, a time or two, when one of their flock moved in to gulp down a bit of effluvium from the ship, I was obliged to abandon birdwatching for a few wretched minutes.

I believe that it was on the 1932 trip—perhaps later—that we found the amputated body of a bald eagle, our national bird. Someone had shot it for the bounty which the Territory of Alaska was then paying for each pair of yellow claws. Eagles were thought to prey on salmon, and I suppose they did take a live one now and then in addition to the hundreds of dead ones upon which they feasted after the spawning season. They were also accused of taking young foxes which might otherwise have matured to be trapped for their valuable pelts. The bounty system prevailed from 1917 through 1952 and brought death to about 125,000 eagles. There were years when, in the remoter hamlets of Alaska, eagle claws were legal tender, accepted as cash by the local merchants. Zoology was then little concerned with finding out exactly what the eagles *did* feed upon. Sufficient it was that certain legislators, to please their fishing and trapping constituents, could perpetuate the bounty system.

But public sentiment for eagles was quietly growing. With the blessing of President Roosevelt, Congress passed the Bald Eagle Protection Act of 1940. Although it contained a clause excepting the Territory of Alaska, that exception was removed in 1959 when Alaska became a state. Today there are about seven thousand nesting pairs of bald eagles in Alaska, more than in the other forty-nine states combined. The greatest remaining threat to their survival is no longer the gun but the chainsaw. Clearcut logging is destroying their huge nesting trees, certain ones of which are known to have been occupied by successive generations of birds for at least a century.

One night about ten o'clock, while we were drifting in Queen Charlotte Sound and taking water samples, hundreds of Leach's petrels—or "Mother Carey's Chickens"—attracted by

APPRENTICE WORK

our floodlights, came aboard ship. They fluttered and crawled over the rigging, talking in harsh, chuckling voices and puking the reddish, oily remains of their last meals of plankton. They were unafraid and could be picked up by hand.

At Ketchikan we visited a game warden who was holding a female wolf in a cage that had a metal floor. Her urine drained into a bottle, to be used later as a lure for trapping the males of her species. It seemed an unfair seduction. I can still see her eyes—black pupils rimmed with greenish gold—burning in the darkness of her cage.

We poked along to Skagway, taking marine samples along the way. Then we turned westward into the open Gulf of Alaska, which was a mistake, for we ran smack into a September gale. The little *Catalyst,* showing over 40 degrees of list on her inclinometer, threatened to capsize. Everyone but the captain was seasick and not a little frightened. My roommate, a young physicist, died after drifting into a diabetic coma. So keenly had he wanted to make the trip that he had not told the leader of his handicap. Sadly, we returned to Seattle, and the *Catalyst* never again ventured far from shore.

Although the 1932 trip was often rough, wet, and miserable, it opened my eyes to the possibility of a career in marine research. And it gave me a glimpse of a new kind of beauty—the clean beauty of ocean wilderness, a beauty that can be felt but never described.

~ In the autumn of 1935 Mary Elizabeth (known as Beth) MacInnes and I were married in a ceremony performed by her father, a Presbyterian minister. He was a Nova Scotian from Baddeck and he had, one summer, tutored the daughters of Alexander Graham Bell. Beth continued to work for many years in social welfare. I have never lost respect for her ability to remember the names and relationships of persons, while she remains impressed by mine to remember the names of bugs, seals, and seaweeds. Although her broad interests do not include zoology, she occasionally reads, for entertainment, the titles of articles in my scientific journals.

"What type," she will ask, "would be writing a 'Bibliography of the Nude Mouse' or 'A Case of Tangled Squirrel Tails'?"

(A zoologist did, in fact, find a cluster of wet squirrels which had crawled into a hollow tree, where they tied themselves in knots and subsequently died.)

When I proposed to Beth she countered, "Do you snore?" She is one who has always seen life with clarity.

~ I seem to have mislaid the summer of 1936, although I do recall that, very tired, I left the University in August with the degree of Ph.D. in zoology. My thesis was entitled "A Limnological Study of Lake Washington, Seattle." Professor Rex J. Robinson added chemical and physical data to the thesis and we published a joint paper in *Ecological Monographs* in 1939.

Limnology, or the study of freshwater streams, rivers, ponds, and lakes, was Professor Kincaid's main research avocation. So, as his student, I drifted into pursuing limnology as a graduate field. It was never to prove directly useful to me in postacademic life and yet the discipline of studying it under an able teacher was surely beneficial. What I learned about scientific methods (researching the literature, using the microscope, keeping accurate notes, and so on) proved helpful in my later research assignments.

Rex Robinson and I made monthly sampling cruises aboard the *Catalyst,* traversing Lake Washington, a body of water covering fifty square miles. We aimed simply to document the physical, chemical, and organic changes taking place in a large, clean lake during the course of four seasons. By chance, though, our report proved to have practical value when the lake subsequently "went bad." It seems that the city of Seattle and a half-dozen suburbs had been pumping increasing volumes of sewage into the lake, thereby adding nutrient chemicals such as phosphates and nitrates. During the 1950s the lake became cloudy and smelly in summer, while the oxygen content of the bottom waters dropped to zero. (We had never found less than 50 percent oxygen saturation.) Having on hand for study a before-and-after picture of the lake's limnology, local government officials created in 1958 a metropolitan district empowered to divert sewage away from the lake. Ten years later, Lake Washington was again a place where hundreds of thousands could enjoy fishing, swimming, boating, and waterskiing.

APPRENTICE WORK

During the course of our investigation it was my special job to study the plankton, that is, the tiny plants and animals (none larger than a match head) which drift or swim feebly in the open waters of the lake. They compose a year-round population, although some are dormant in winter. Peering through the microscope, I listed 107 species of plants and animals. Some were beautiful—encased in crystal shells and filled with yellow pearls of oil and green plates of chlorophyll. Others whirled or jerked along their aimless courses, propelled by movable hairs, whips, or paddles. I barely escaped thinking of them as individuals with personalities. At any rate, I saw them as wondrous particles of life, complete and self-renewing, extending through a transparent system as strange as the outer dimensions of space.

In September 1936 Frank Brockman kindly let me return to Mount Rainier to organize the park museum's "lantern slide" (projection slide) collection. Modern color-reversal transparencies, such as Kodaslides, had not yet been perfected. We used glass-plate photographs tinted by hand with transparent dyes which simulated the colors of nature.

Meanwhile, I looked for steady work. Many of us who had graduated during the Depression hoped that one government agency or another would eventually hire us. There were virtually no openings for zoologists in private industry. From 1930 to 1936 I took seven civil-service examinations in a variety of fields.

2

The Aleutian Expedition

IN April 1937 I was appointed junior biologist, salary two thousand dollars a year, in the United States Bureau of Biological Survey. The central offices of the Bureau were in Washington, D.C.; I was to be stationed in Olympia, the capital of Washington State. I had now entered a career in civil service that was to end thirty-two years later. My father had just retired from the Bureau at age seventy and I'm sure that his performance record in that agency was reviewed in the city of Washington while his son's application was being weighed.

The Bureau was then nearing the end of its political life. It was a Department of Agriculture agency which had been created in 1885 as the Section of Economic Ornithology with responsibility for investigating the good and bad effects of birds upon agriculture. Under its first chief—Merriam—the Bureau grew and diversified. It took on responsibility for exploring, describing, and mapping the largely unknown wild bird and mammal faunas of the United States. It launched a serial publication, *North American Fauna,* which is still being published although the Bureau itself closed its doors in 1940. *Fauna* No. 1 (1889) mapped the species of North American pocket mice, while the most recent issue (1975) describes the life history and the ecology of the screech owl in northern Ohio.

Nearly a year went by before I was destined to meet my supervisor, Hartley H. T. Jackson (known from his initials as Alphabet Jackson), who maintained an office in the U.S. National Museum of Natural History, in the city of Washington. He was a person for whom the label "gentleman" seemed exactly right. Mild of manner, articulate, and scholarly, he was able to drop zoology whenever his flower garden called for attention. He directed me by mail or telegram—long-distance telephone calls then being prohibitive. As a green, bewildered young government worker I found the lack of personal contact frustrating.

He started me off with a pair of binoculars, a camera, and a thick administrative manual known as the Black Book, which detailed various sins of omission and commission. Most of the proscribed acts were those any decent employee would avoid intuitively without benefit of a guide. One warning was against writing "petulant" letters to Washington. I should like to think that mine were never petulant, although some were surely intemperate.

The chief of the Bureau in 1937 was Ira N. Gabrielson, a big, jolly, keen-eyed man who later received three honorary doctorate degrees, was author of a half-dozen wildlife books, and held many administrative positions. Although "Gabe" operated at government levels far above mine he was always ready to chat with me or with any other young biologist in from the field for a visit to Washington. He himself had spent his happiest years roaming through the West. Wildlife management as a profession originated in 1937; Gabrielson died in 1977, in its fortieth year.

Aboard the Brown Bear

While the ink was scarcely dry on my appointment I was directed by Hartley Jackson to join the ongoing Aleutian Islands Expedition (1936–38) under Olaus Johan Murie. So feebly developed was my sense of history that I did not then appreciate what a rich prize and privilege the assignment was. Our chief mission was to make a biological survey—or "wildlife inventory," in the words of the assignment—of those treeless, nearly unpopulated islands that reach for eleven hundred miles westward from the Alaska Peninsula. The expedition was financed by what Olaus called "duck money." Thanks to the influence of sport hunters, federal funds for the study of waterfowl nesting grounds in North America were plentiful, whereas funds for purely exploratory research were not. We were soon to find that few ducks or geese do, in fact, nest in the Aleutians, but, nonetheless, we spent the expedition's money, untroubled as to its source.

Although more than seventy of the Aleutian Islands, occupying more than two million acres, had been set aside in 1913 as a national wildlife refuge, almost nothing was known of their fauna and flora at the time our expedition was launched. Besides fox trappers, few men had chosen to land on those bleak stepping stones to Asia. Lying as they do between the cold Bering Sea and the warmer North Pacific Ocean, the Aleutians are a region of treacherous currents, persistent fogs, and sudden

2. Leo K. Couch (*left*) and Ira N. Gabrielson (*right*), zoologists of the U.S. Bureau of Biological Survey, photograph alpine flowers in Mount Rainier National Park, 1933.

THE ALEUTIAN EXPEDITION

winds known as "williwaws." Thanks, however, to the skill of her captain, our expedition ship, the *Brown Bear,* threaded the islands for three summers without mishap.

Our modus operandi was to anchor off an island, go ashore by dory, then spend the day observing, photographing, and collecting plants and birds. Most of the islands supported no mammals at all. Some were populated by introduced blue foxes (planted to become fur-bearer stocks) and tundra squirrels (brought in as fox food), or by house rats escaped from ships and subsequently gone feral. We hiked for many a mile on tundras rich in green vegetation and suitable shelter, yet saw no signs of shrew, mouse, rabbit, or other mammal.

~ At times even now when the wind moans at the corner of my studio and the rain slashes against the window I am back on an Aleutian island, moving up a grassy slope, feeling wet ryegrass whipping my rubber boots, and tasting salt spume on my lips. All alone. The others of the party are exploring other parts of the island. In the lee of a volcanic boulder shaped like an Easter Island figure a winter wren pours out its bubbling melody, ignoring me as I pass. I am the first and the only human it will ever see. I fumble in my windbreaker to write "wren" on a sheet of waterproof paper, to remind myself—come evening and the shelter of the *Brown Bear*—to record the presence of *Troglodytes troglodytes* on this island on this date.

As I near the backbone of the island, the knee-deep grass gives way to pumice fields dotted with dwarf willow and crowberry. I chamber a shell in my .410 shotgun, hoping to surprise a rock ptarmigan. And so I do, but before I can drop the bird it explodes into flight and falls on a swift, curving trajectory down the mountainside, crying alarm in a rattling voice. This bird is a very different actor from the placid white-tailed ptarmigan I had known on Mount Rainier.

After dinner aboard ship, I open my pack, press plants between large sheets of blotting paper, preserve the eggs of a rosy finch, label vials of plankton collected in a lake, pickle some beetles for Mel Hatch (the Pacific Northwest expert on these insects), and enter in my photo catalog a record of the pictures taken that day. Then up to the galley for a nightcap of hot

3. Author takes a plankton sample from a lake on Semisopochnoi Island, Aleutian Islands, 1938.

tomato juice stirred with canned milk, and to bed. When I wake in the morning, the *Brown Bear* may be anchored off another island.

~ The kind of zoology we pursued during the expedition was mainly exploratory. It resembled the zoology of Lewis and Clark and, later, of Merriam's field crews, in that it placed great importance on the collecting, identifying, and cataloging of specimens. The geographic ranges of many plants and animals had yet to be mapped (and, with respect to the rarer forms, remain so today). We collected hundreds of birds and prepared their skins to be studied later in the National Museum by Murie and other specialists. To prepare a bird skin (or, in laymen's lan-

THE ALEUTIAN EXPEDITION

guage, to "stuff" it), one removes the fleshy parts of the body, scrapes the fat from the skin, dusts the skin with arsenic to protect it from moths, and substitutes for the fleshy parts a torso made of rolled cotton, hemp, or excelsior. The wings and feet are folded in a prescribed position and the specimen is wrapped in a cheesecloth shroud to dry. A label written in permanent ink and tied to one leg records certain measurements taken while the bird was fresh, as well as the date, place, and collector's name.

I take the same pleasure from handling a label signed by a famous ornithologist of the nineteenth century as a bibliophile would take from finding "A. Lincoln" inscribed on the flyleaf of an old book.

We did not land on the smaller Aleutian islands—many of them unnamed—but, rather, circled them in a dory powered by an outboard motor. One man ran the motor and steered, another sat in the bow watching for submerged rocks and kelp beds, while a third scanned the shores and the shallows for seabirds, harbor seals, sea lions, killer whales, and other features of interest. The work was devilishly cold and wet, and not a little dangerous. At the first sign of an incoming fog bank we headed for the *Brown Bear* or in the general direction of her whistle.

Aboard the ship anchored at night we often lowered a baited, wire-screen cage to attract crabs, starfish, sea snails, and other specimens. Soon after I joined the expedition in 1937 Murie realized that I knew nothing about Alaskan birds and mammals but did know a bit about marine and freshwater organisms. So he suggested that I collect, photograph, and take life-history notes on the invertebrates and fishes that we would be encountering. These he termed "the supporting fauna" for the larger animals, including the birds, foxes, and sea otters, which were the principal objects of our faunal investigations.

I was repeatedly impressed by the great volume of invertebrate life in those cold waters. At times, a full quart of amphipods (or sand fleas) would pour from the crevices of the bait we had lowered into the sea. Whenever we had the carcass of a fish, bird, or mammal to be skeletonized for study we recruited the help of the amphipods. They would pick the bones clean in a night or two.

Invariably, when the *Brown Bear* was anchored in a sheltered cove waiting out a storm, the ship's crew would start fishing for cod. They used a method known as jigging, in which naked hooks attached to a shiny strip of brass at the end of a line are rhythmically jerked upward. I would open the stomachs of the fish to search for invertebrate specimens. Surprisingly, out of the stomach of a two-foot cod taken off Chuginadak Island there popped the fresh head of a cormorant; out of a cod from the shallows of Ogliuga Island, the whole body of a paroquet auklet. I thought these finds remarkable until I later learned that diving birds are not uncommonly seized by cod. Perhaps I should not have been surprised, for I once opened a cod stomach containing a turkey foot, a boiled potato, and a penny matchbox! That individual had been lurking beneath our ship, waiting for manna to descend.

Several times while fishing in the kelp beds we caught rock greenlings, a species having green flesh and viscera the color of mint ice cream. Although the green fades to gray in the frying pan, it is startling to one who sees it for the first time. I haven't the faintest idea why the inside of a fish should be green.

Birds, Foxes, and Sea Otters

In August 1938 we landed with keen anticipation on the famous "disappearing island" of Bogoslof, a black lava mass jutting like a fist from the Bering Sea. It was said to rise and fall and to move its position. That notion originated in faulty navigation charts and in a certain degree of real, though slight, shifting of the mass as a result of vulcanism. At the time of our visit Bogoslof had been colonized—without help from man—by sea lions, murres, puffins, gulls, and a few cormorants. Isolated patches of grass had sprouted from seeds blown in by fierce Aleutian gales. Some of the island's rocks were still warm to the touch.

When I returned to Bogoslof thirty years later, in 1968, the whole central plateau was now a luxuriant field of green and the rocks were cold. I was puzzled at seeing the torn, bloody bodies of several puffins, for I had thought the island remote from

THE ALEUTIAN EXPEDITION

predatory animals. Then I heard a furious scream and I looked up to see a bald eagle circling high. A pair of eagles had found their way to Bogoslof and had built their nest on a rocky pinnacle. Here, among sea-bird colonies, surrounded by food, their future seemed bright indeed.

I shall be interested in following the history of Bogoslof to learn the outcome of this, one of Nature's experiments in eagle and prey interactions. The most likely outcome is that only two or three pairs of eagles will persist as breeding residents of the islet, while subadult birds will be forced to emigrate. The limiting factor in population growth will not be food or nesting space but psychological space. The adults, motivated by their territorial instinct, will begin to feel crowded and will haze the youngsters out of the ecosystem.

On a dismal, gray day while our ship was cruising through Unimak Pass the captain called me to the wheelhouse at five-thirty in the morning. I saw that we were passing through a vast concentration of whalebirds, or slender-billed shearwaters, covering an area that I later estimated as being two miles wide and fifty miles long. It was a display, said the captain, unrivaled in his long experience in Alaska. To me it was a demonstration of the incredible richness of the marine pastures of the North Pacific. The birds were feeding on plankton, building up fat stores for their long migration to Australia, where, as "mutton birds" in down-under parlance, they would soon be nesting.

Just as I had seen petrels board the *Catalyst* in 1932, I saw them visit the *Brown Bear* on an August night in 1938 as she lay off Amchitka Island. On this occasion, only a few were Leach's; most were fork-tailed. (Zoologist Joseph Grinnell once said that while he was camped on an island in Sitka Bay he was unable to keep a fire alight because the petrels flew into it in such numbers that they extinguished it!)

The Aleutian Chain was set aside in 1913, not only as a preserve for native birds but as a self-perpetuating fox ranch. Between 1913 and 1932, foxes were planted on sixty-eight of the islands. No one in charge of the operation seemed to know that birds and foxes would be unlikely to coexist in peace and harmony. Foxes were already living on the big islands of Attu and Agattu, seeded there in prehistoric time, one supposes, by

4. A Leach's petrel, confused by the *Brown Bear*'s lights, rests on the shoulder of a ship's visitor, Alastair Macbain, 1938.

individuals carried on drifting sea ice. Foxes did not reach Buldir Island, and only on Buldir did the Aleutian Canada goose, a relict form, survive to the present day.

It was one of our missions during the expedition to estimate for each island whether its foxes were subsisting mainly on beach life (such as sand fleas and stranded fishes) or on birds (such as murres, petrels, auklets, gulls, and ptarmigans). With that objective in mind we picked up fox droppings, or "scats," and wrapped them in pages torn from pulp magazines. We later examined the scats and made a rough tally of their contents.

"I signed on as a zoologist," moaned Doug Gray, a member of our party, "and I find myself picking up after foxes."

We recommended that the foxes be removed from those islands where bird remains were dominant in the scats. The Fish and Wildlife Service (successor to the Biological Survey) began to act on our recommendation but soon found that getting rid of foxes is easier carried out on paper than in the wild. Its agents air-dropped the deadly, tasteless poison, Compound 1080, on Amchitka Island. It killed the foxes . . . and the native ravens as well. The agents turned to M-44 "coyote gitters" (explosive cyanide devices) but found them ineffective. My latest information is that they are now using steel traps and rifles.

THE ALEUTIAN EXPEDITION

The introduction of foxes between 1913 and 1932, evidently without concern for their future impact on the environment, had been a mistake. It was, however, an action typical of a time in the early development of wildlife management when animals were seen essentially as inert resources, or "rawstuffs." To now correct that old mistake it was necessary to kill off the foxes which were living on islands where they should never have been planted. Two of the killing methods—poison and the leghold trap—subsequently used by the government were inhumane. I condemn them now for their cruelty, although years ago when I was younger I would have thought mainly of their efficiency.

How to get rid of unwanted, or "weed," species is one of the more difficult questions posed to wildlife management zoologists. The costs of doing so are often high, as in New Zealand, where the descendants of imported deer are being pursued throughout the year by squads of government marksmen. Moreover, few zoologists like to kill animals, even when the killing is justifiable. In the example of the Aleutian foxes, I suppose that the animals might best be eliminated in winter, when they are hungriest, by shooting them either in live traps or along their trails. Although this proposal may sound callous, it is in fact humane.

~ I recall a day on Tanaga Island when, hiking across the tundra, I stopped to investigate the abandoned shack of a fox trapper. The wind slammed the door behind me. On the floor I saw the dry bodies of two rosy finches which had entered through cracks and had not found their way out. I saw on a shelf a rusty phonograph and a record that carried me back in a flash to childhood. It was labeled "Lavinsky at the Wedding"—a dialect piece thought screamingly funny in its time. I mused . . . what pleasure could a lonely trapper take from hearing that voice repeated a hundred times? But I suppose that any human sound at all would be a welcome respite from the moaning of the sea wind and the sad cries of gulls.

~ Murie was especially interested in the Aleutian sea otters. Their numbers, tragically depleted by over-hunting in the last century, had risen from near zero in the early 1900s to perhaps five to ten thousand in the 1930s. Throughout Alaska they were rigidly protected (although a person who found one dead on the beach might salvage it, have its pelt certified by a game warden, and keep it).

Hartley Jackson had told Murie of the National Museum's desperate need for a sea otter specimen, so Olaus obligingly "found" a splendid eighty-pound male, its body still warm. Later, when its skin and skull arrived in Washington, Jackson realized that the Museum's faded old display specimen had been mounted with its right and left hind legs reversed! Millions of visitors had seen it and none had caught the mistake.

At the time of our survey, about three thousand otters were thriving in the shallows of Amchitka Island. The island was, in zoological terminology, the metropolis of *Enhydra lutris*. Because the Coast Guard had seen strange men walking on the beach there and had supposed them to be poachers, the Bureau of Fisheries had built a small patrol cabin on Amchitka in 1937. We visited it in 1938 and met two pleasant young men who had recently been hired as sea otter wardens. They told us in suppressed excitement that they were holding a baby sea otter captive on the far side of the island, feeding it on canned milk. After tramping several miles across the rough tundra to see this first for science, we had to tell the wardens gently that their charge was a young harbor seal!

Last of the Whaling Stations

At Akutan Island in 1937 and 1938 we visited a whaling station where fin, humpback, sperm, and blue whales were being processed commercially. The *Brown Bear* anchored one night alongside the station. At dawn I was awakened in my cabin by a most horrible smell. I went on deck to find that workmen had punctured the belly of a humpback whale, deflating it before hauling it up the slip and onto the butchering platform. The gases of decomposition were so sulfurous that they turned the white paintwork of our ship light brown—a discoloration which the crew later removed with strong soap and water.

THE ALEUTIAN EXPEDITION

At Akutan I first looked a whale straight in the eye. In memory I stand again on the slippery dock, watching the workmen peel the blubber from the forty-foot body. White fat falls from red flesh, steaming in the chilly air. The butchers' knives go deeper and deeper into the frame. I look with disbelief at the animal's heart (the size of a kitchen stove), at the pink, shrimplike food pouring from the ballooning stomach, and at the twelve-foot arm or flipper rising stiffly toward the sky. The dark eye rimmed with white stares at a far-off place I cannot see. ... A dead beast in a dying industry. When the Akutan station closed at the end of 1939, Alaskan shore whaling was finished.

The other Alaskan stations had already folded—Tyee in 1910, Whale Bay in 1912, Port Armstrong in 1922, and Port Hobron in 1937. They ran out of whales. Today, the zoologists who study whales in various parts of the world are turning to the history of those and other shore-whaling stations for insight into the migratory habits of whales. It seems clear that the pathways

5. Author (*left*) watches a sperm whale being butchered at Akutan Island, Alaska, 1937.

of the whales through the sea are deeply ingrained in behavior and, consequently, that commercial whaling can quickly reduce local populations to levels from which they may not be able to rise. (I shall write more about this later; sufficient now to note that, as recently as 1978, the Cheynes Beach, Australia, shore station closed after having processed more than fourteen thousand sperm whales.)

While we were at Akutan in 1937 we met a Coast Guard inspector who had been assigned to enforce the new Whaling Treaty Act of 1936. He was unhappy. Undersize whales were being killed and he saw no feasible way to take the evidence—if by "evidence" was meant the whale itself—to the nearest federal commissioner. He and the whaling superintendent were barely on speaking terms. Although their desks were only a few feet apart they were communicating mostly by letter.

Later, the superintendent led us to a shed where, at the request of Remington Kellogg, he was storing samples of whale stomach contents. These would eventually provide zoological evidence of the summer diet of whales. As we approached the shed upwind my nostrils began to twitch. Within it, on wooden shelves, were scores of glass fruit jars containing the samples—brownish or pinkish matter in various stages of decay. The superintendent had not clearly understood Kellogg's instructions.

"I leave the jar lids off," he told us, "until the stuff stops *working*, then I add the formalin preservative."

But in spite of his crude technique some of the samples were later useful; their contents were identified at the University of Washington as plankton organisms, squids, and fishes.

~ Nearly forty years later I was privileged to watch a Bryde's whale being butchered at Onagawa, Japan, and to marvel at how quickly and cleanly this operation proceeded in contrast to the earlier one at Akutan. In the company of Akito Kawamura, of Tokyo's Whales Research Institute, I was traveling in Japan in 1976 on an assignment for the National Geographic Society. As we stood on the whaling platform in early evening, a catcher boat towing a forty-one-foot whale came in from the open sea. Under floodlights, a score of yel-

THE ALEUTIAN EXPEDITION

low-helmeted workmen attacked the carcass with knives, power saws, and winches. Its bulk melted rapidly and, an hour later, the men were hosing down the platform. The whale's great mass had become neat piles of meat, salted or iced for human food, and piles of bone and waste destined to become animal feed or fertilizer.

Akito supervised the collecting of the whale's ear plugs—each a small, waxy body shaped like a golf tee—by means of which he would later estimate the whale's age. I was struck by the combination of brutal machine power and delicate finger power which he used in collecting the plugs. First, the whale's head was torn apart by steam winches, to the sound of popping cartilages and tendons. A man with a flensing knife bared the inner ears and then Akito, with tweezers, lifted out each plug and plopped it into a vial of formalin.

Although, for more than a thousand years, the Japanese have used small amounts of whale meat as a protein food for humans and livestock I believe they ought to cease doing so. Whaling in the world ocean is increasingly more difficult to regulate. And it is cruel; the standard killing weapon is a fragmentation bomb. The United States Government first called in 1972 for a worldwide ban on whaling and, at last report, had not changed its position with respect to the need for one.

Sorting Out the Collections

Murie did not return with the staff of the *Brown Bear* to the Aleutians in 1938—the third and last year of the expedition—but gave us advice before we sailed. In the meanwhile, I had established (in March 1938) headquarters in the University of Washington's forestry building, Anderson Hall. That was destined to be the first of a series of sea-mammal research headquarters, each of which I will refer to as "the Seattle laboratory." Since 1957, the laboratory has been situated on Navy land a few miles from the University, where it has grown enormously.

I returned on the *Brown Bear* to Seattle in the fall of 1938 and began to sort out the jumble of specimens—dried, pressed, or in liquid—collected in 1937–38. They included marine, land, and freshwater invertebrates, fishes, seaweeds, and land plants. Several were new to science. Murie's name is now immortalized in a marine snail, *Anabathron muriei,* collected from a sea-otter scat, while mine will live forever in a mud clam, *Liocyma schefferi* —a plain little thing barely half an inch long. A year after his death, the Murie Islets, a group of rocks near Simeonof Island, were named in honor of the man who contributed so richly to knowledge of the Aleutian fauna.

Most of my specimens were identified by experts in the National Museum. I had naively supposed that those men and women would drop whatever they were doing in order to study *my* collection. I did not know of the pressures under which they continually work.

"If you wait your turn," said Waldo L. Schmitt, curator of invertebrates, "you'll wait fourteen years." Then that generous soul began at once to devote part of his time to helping me.

Was It All Worthwhile?

The manuscript report of the Aleutian Expedition wound its way slowly toward publication. Twenty years later—and after World War II had intervened—it appeared, in 1959.

The most important findings of the expedition were those which may be called *baseline data.* Time and again in the present Age of Ecology, land-use planners and wildlife managers search in vain for facts about pristine conditions in this or that ecosystem. What was the system like, they ask, before civilization disrupted it? What was it like when it was in natural equilibrium?

We had the historic privilege of describing, for example, the ptarmigan of Amchitka Island before house rats, introduced accidentally with Army goods, destroyed the last bird. And we recorded a stage in the natural progression, or emigration, of Asian bird species toward the American mainland. Among these were the common teal, the Aleutian tern, and the whiskered auklet. We learned the Aleut native names for many birds and mammals from a people whose culture was already crumbling and is now nearly gone.

Our baseline data were taken in an essentially precontaminant period, that is, pre-DDT, pre-atomic fallout, and pre-coastal oil exploration. Zoologists of the future, studying the sea birds and sea mammals of the Aleutians, will, I believe, be grateful for the data—especially on population numbers and geographic distribution—that we were able to record in the thirties.

Some readers of the expedition report find interest in the life-history details of individual species. They learn, for example, that blue foxes eat more than sixty kinds of plant and animal foods ranging from mosses and cranberries to human skin (Aleut mummies). They learn that the Bergmann principle, or the tendency of birds and mammals in cold regions to be larger than their nearest relatives in warmer ones, is supported by measurements of certain Aleutian species. Thus, "giantism" is evident in the song sparrows, Savannah sparrows, and rosy finches, and in the huge brown bears of the Alaska Peninsula. (It is a consequence of the fact that a large animal can withstand cold better than can a small one.)

The Aleutian Islands Then and Now

Before World War II, fewer than five hundred people lived in the Aleutians. Now Adak Island alone supports nearly four thousand Navy persons and their dependents. Amchitka Island, then a virgin wilderness, now wears the scars of the atomic (nuclear) bomb tests of 1965 (Long Shot) and 1971 (Cannikin). Scarring of the fragile tundra occurred before the tests, during the construction of roads, wharves, and buildings. No one involved at the time seemed to care that Amchitka was a national wildlife refuge and had been one since 1913. Writing in *Audubon* Magazine, George Laycock mourned in 1972 that "One cannot travel to Amchitka without being offended by man's impact and his willingness to leave behind what he no longer wants. . . . There should be no such thing as a throwaway island."

ADVENTURES OF A ZOOLOGIST

When in 1937 we saw Attu Village on Attu Island, about forty families lived there. Its frame houses, a school, and a Russian Orthodox church rested in the shadow of a great mountain still covered with snow in early summer. In the village we watched an old woman with a beautifully wrinkled face weaving one of the now priceless Attu baskets, using only the middle fibers from ryegrass *(Elymus)* blades. The village chief, Mike Hodikoff, spoke the Attu names of the local birds as Murie transcribed them in his notebook.

In 1942 Japanese invasion forces took all the Attu people captive. About twenty-five, including those born in Japanese prisons, were repatriated after the war. Although the survivors wanted to return to Attu they were landed at Atka, five hundred miles nearer the mainland of Alaska. (Perhaps for reasons of security?) I visited Attu shortly after the war and stood among the rubble of burned and shattered homes, reading the words on a monument erected to a people who never came back.

During the first summer of the expedition (1936), before I had joined it, the *Brown Bear* party had stopped at Kagamil Island to investigate a white fumarole shooting steam high into the air. Nearby, in a warm, dry cave, Murie found an ancient cemetery of Aleut mummies wrapped in grass matting and sea otter skins. He radioed the news to ethnologist Ales Hrdlicka, who was then on a dig in the Aleutians and who, later in the summer, collected about fifty mummies to take back to the National Museum.

My only meeting with the great Hrdlicka was brief. His ship and ours chanced to lie at anchor in the Bay of Waterfalls in 1937. He came aboard the *Brown Bear* for dinner and I proudly showed him an artifact I had dug from the sands of Little Kiska Island—a graceful platter carved from the intervertebral disk of a whale.

"You have no use for this?" he inquired softly, lifting the relic from my hands. And of course I hadn't. Relic hunting for the sake of souvenirs has always (and rightly) annoyed professional students of the ancient past.

Out in the Aleutians where the sea wind stirs the rubble of long deserted homes and where brown bones fall from crumbling earth one feels very close to history.

THE ALEUTIAN EXPEDITION

Olaus Murie

I first met Murie in 1937 a few days before the *Brown Bear* left Seattle for the Aleutians. He quickly became for me the ideal zoologist. Of blue-eyed Norwegian stock, he spoke softly but surely. He wasted few motions; when he put his foot down there was a place for it. Whatever he did seemed so *right*. Naturalist, author, artist, philosopher; charter member of the American Society of Mammalogists, the Wildlife Society, and the Wilderness Society—he was all of those. It is a sign of his clear vision that he saw mankind's unending need for wilderness. He understood that wilderness is more than geography, it is a place in the human mind.

Looking back down the years I can see Olaus sitting in the cabin of the *Brown Bear*. He is preparing a study specimen of a golden sandpiper chick scarcely larger than a bumblebee, talking to its feathers, persuading them to lie softly on the cotton body.

"If a feather offend thee, pluck it out," he tells me with a smile as he watches me trying to copy his technique.

Only once did I ever see him shattered—after a young zoologist in our party had lost his way on remote Agattu Island. At the end of several days' search, Murie radioed to the man's wife that there was little hope of finding him alive. At that point, bone-tired, and tormented by what he felt was his failure as a leader, Olaus reached for a shotgun and dropped a white gull from the sky. For a man whose everyday life defined humaneness that act was a *cri du coeur*. (The lost zoologist, unharmed though hungry and wild of eye, eventually found his way back to the *Brown Bear*'s anchorage.)

In 1959, at age seventy, Murie was featured in *Life* Magazine and was quoted as saying, "The out-of-doors is our true home, and being there gives us solitude and leisure to speak to ourselves and not to others. When we speak to ourselves, we are apt to be more honest."

I was pleased to be asked by Murie's widow, Margaret, to write a foreword to his 1973 posthumous book, *Journeys to the Far North*. The foreword closes with these words: "When I last talked to Olaus, he spoke in distress of the Computer Age,

6. H. Douglas Gray (*left*) and Olaus J. Murie (*right*) prepare museum specimens of birds aboard the *Brown Bear* in the Aleutian Islands, 1937.

THE ALEUTIAN EXPEDITION

artificiality, and of man's abuse of the wild places of earth. I said, smiling, that he was now an ecologist and ought to be happy with his new title. He wrinkled his nose and said 'Gee!' In his heart he had not changed. The truths he recognized early and spent a lifetime shaping into words and pictures were still the same old truths. He knew it, and I knew it."

3

Studies of Land Animals

IN early 1938 I took train a to the city of Washington to meet Hartley Jackson for the first time and to find out what my next assignment was to be. I learned that Jackson was in favor of launching a biological survey of the mammals of Washington State while his assistant, Leo K. Couch, thought that I should study the status of endangered fur-bearers, especially marten, otter, wolverine, and fisher, in the Pacific Northwest. As a consequence I tackled both assignments more or less simultaneously while also winding up the Aleutian studies. Two years later I was sent to the Pribilof Islands on the first of many assignments which were gradually to lead me away from the land and to the sea.

The Pacific Northwest fur-bearer study eventually fizzled out. It was replaced by a more ambitious study carried out by the new (1932) Washington State Game Department. The survey of the mammals of Washington was completed by Walter W. Dalquest, an undergraduate student at the University of Washington and graduate student at the Universities of California and Kansas. Published in 1948 by the University of Kansas, it was his doctoral thesis. Walt and I worked together sporadically for several years before it became evident that I would not be able to carry my fair share of the load as collaborator and coauthor. (He now teaches at Midwestern University at Wichita Falls, Texas, and has turned to research on fossil mammals.)

We keenly enjoyed our collecting trips in the deserts, forests, and mountains of Washington. Walt was a lively companion—brilliant, imaginative, and friendly, with a charming disrespect for tradition and rules. Had he been born a generation later I suppose he would have been called a hippie.

I recall our excitement when we drove into Pend Oreille County, where the Rocky Mountains barely cross the northeast corner of the state, and found mammal species new to us. The ground squirrels and marmots, among others, are elements of the Rocky Mountain fauna which differ from their relatives in the coastal Cascade and Olympic Mountains. By surviving on trap bait (dry prunes, dry oatmeal, and peanut butter washed down with water) we were able to stay a day longer in that fascinating place than we had planned.

Elsewhere in the state we discovered and described several new races of mammals. A shrew living among the spray-wet vegetation of Destruction Island, off the Olympic seacoast, now carries the name of *Sorex trowbridgi destructioni*—a name nearly as long as its body. We found in the soils of the Yakima Valley a pale-furred mole which seemed to deserve its own name, so we dubbed it *Scapanus orarius yakimensis.* On certain grassy prairies among the evergreen forests of southwestern Washington we discovered three new races of pocket gophers, each of which had evolved since the Ice Age within the confines of its own small prairie. The discovery shed light on a matter of intense interest to evolutionists, namely the *rate* of evolution, for in this case subspecific differences must have appeared in less than twelve thousand years (or twelve thousand generations).

The Mysterious Mima Mounds

During the collecting years with Walt I developed a special interest in the so-called pimpled plains of the American West and their type locality, Mima Prairie, near Olympia, Washington. This prairie is dotted with peculiar earth mounds created in postglacial time by gophers *(Thomomys)*. Numbering in the tens of thousands, the mounds resemble great eggs partly buried in the earth, some of them higher than a man's head and holding fifty cubic yards of fine black soil. They share with beaver dams the distinction of being the most impressive structures

7. Mima Prairie, Washington, 1966, is dotted with huge earth mounds up to 7 feet high, built by gophers on a glacial outwash plain.

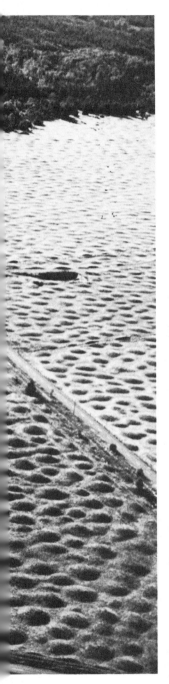

8. A mountain beaver, a "living fossil," at the entrance to its burrow, 1967.

built by any mammals on earth. Because they rise from the outwash gravels and silts of the last continental ice sheet, the mounds were at first thought to be products of moving ice or of freezing and thawing.

By coincidence, Walt and I were trapping gophers on the mound prairies near Mima while he was taking a course in geology. In a flash of intuition he saw that each of the huge mounds must be the territory or nesting center of a gopher family. He invited me to help him search for proof, and by 1942 we had gathered enough evidence to impress the editor of the *Journal of Geology,* J. Harlan Bretz, who himself had studied the Mima mounds some thirty years earlier.

After reading our manuscript, Bretz tactfully suggested that we postulate the origin of the mounds by *ice,* the gophers subsequently adopting these neat "prefab" piles of earth as their homes. We countered that no widely accepted theory of mound formation by ice had yet been put forward (nor has one to this day), and that we rather liked the simplicity of the gopher theory. Bretz was a fair editor; he did not press his point.

Examples of the Mima microrelief, instantly recognizable by one who has studied them, have now been mapped on grasslands in the vast region bounded by British Columbia, Minnesota, Louisiana, and southern California. Scores of articles about the mounds have been written, and in 1966 Mima Prairie was designated by the National Park Service as a Registered National Landmark.

I should add that gopher origin of the mounds is not universally accepted. Among the skeptics are geologists who have not studied the mounds in the American Southwest, far south of any known influence of ice. And one holdout is a farmer living near Mima with whom I recently talked. "Well," he said thoughtfully, "they must have been mighty big gophers." But they were not. They were look-alikes of the four-ounce rodents which even now are busily tunneling through the topsoils of the pimpled plains of the West.

STUDIES OF LAND ANIMALS

The Mountain Beaver, a Living Fossil

In the 1930s I developed an interest in another small rodent of the Pacific Northwest—the mountain beaver *(Aplodontia)*. Belying its name, it is rare in the mountains and is not at all like a beaver. Weighing two or three pounds, it is a shy, nearly voiceless, dingy brown animal that spends most of its life in a labyrinth of gloomy underground tunnels. It inhabits the damper forested regions of Washington, Oregon, California, and Nevada—and nowhere else in the world. The most primitive of all living rodents, it is the only survivor of a stock including about thirty known fossil species—the oldest dating back sixty million years.

My father, too, had been interested in mountain beavers. He began to study their habits when the Biological Survey assigned him in 1914 to deal with the injurious birds and mammals of Washington. During his investigations he collected several gigantic mountain-beaver fleas, one of them more than three-eighths of an inch long. It proved to be a new species and was named *Histricopsylla schefferi*. Father was fond of saying that he would be remembered longest for having discovered the largest flea in the world!

When, later, I moved from Seattle to an acre of wild land in suburban Bellevue I had to oust the resident mountain beavers to protect new ornamental plantings. Our son, Brian, then in high school, trapped the pests at twenty-five cents a head. Consequently, he jumped with excitement when I received a letter from a California zoologist offering three dollars apiece for 150 specimens. Brian set out a trapline and followed it faithfully once a month in rain, snow, or fair weather. Often in the darkness of winter he had to grope by flashlight through tangles of wet fern, salal, Oregon grape, thimbleberry, and trailing blackberry.

Still later—in fact, up to 1970—I live-trapped and supplied mountain beavers to North American zoologists who wished to study the characteristics of the beast. My modest, bring-'em-back-alive business was never intended as a money maker, although I did charge twenty-five dollars a head.

The mountain beaver's water balance and kidney structure were of special interest to Egbert W. Pfeiffer, at the University of Montana. He found that the animal needs far more water than does the average rodent. Zoologists had experienced poor success in keeping mountain beavers in captivity and had supposed that they die of shock. In reality, what they need is plenty of water.

Michael M. Merzenich, at the University of Wisconsin, found that the ear of the mountain beaver is ultrasensitive to low-pitched vibrations. Because the animal has no weapons other than its teeth, its acute hearing presumably enables it to detect an approaching enemy in time to "freeze"—and thus escape notice.

Tracking a Lost Herd of Deer

In the winter of 1939–40 Dalquest and I followed to its source a rumor that white-tailed deer were still breeding along the lower Columbia River where Lewis and Clark had found them in 1806. We visited the site, and indeed found what the local farmers were calling "tideland deer" or "cottontail deer." As we approached a pasture in the falling light of day we surprised two yearlings and watched them curtsy under the lowest strand of a barbed-wire fence only twelve inches above the ground. They seemed miniatures. Each raised its tail in alarm, displaying the white "flag" that distinguishes the white-tail from the common and abundant black-tail deer of the American West. We followed many deer trails, often dropping to our knees to penetrate a thicket of willow or osier dogwood. In the face of competition from black-tails, the white-tails have evidently been selected for survival by the advantage of their smaller bodies, which enable them to forage beneath cover that would discourage a black-tail.

To authenticate zoologically our rediscovery of the race, we collected twenty-three sets of antlers and sent them to the National Museum. The antler tines of a white-tail grow from a single beam, while those of a black-tail sprout from a two-forked beam. We picked up most of the antlers in the woods; a few were given to us by farmers who had saved them as trophies. And we shot an eighty-eight-pound doe whose remains were to become the first recorded skin-and-skull, complete with field measurements, of the Columbian white-tail race.

We estimated that the relict herd numbered 100–200 animals on the Oregon side of the river and 400–500 on the Washington side. In 1940 the *Journal of Mammalogy* published an account of our investigation, and in 1972 the Fish and Wildlife Service established the Columbian White-tailed Deer National Wildlife Refuge, comprising over five thousand acres of islands and mainland. We had asked in our 1940 report, "What can be done to preserve the race . . . and at the same time protect the farmers whose interests in that region are steadily increasing?" A time lag of thirty-two years between the question and its answer may seem long, but it isn't unusual. Land-use issues are rarely settled quickly.

Last of the Sea-otter Hunters

Continuing my survey of Pacific Northwest animals, I made a brief side excursion into history. It seems that the sea otters of the Washington coast had been exterminated by hunters around 1910. Thirty years later it was my luck to meet Clara L. Minard, an elderly woman who as a girl had watched the animals playing in the surf and had actually photographed, before the turn of the century, sea-otter pelts brought to her father's store near Oyhut. She let me copy the photographs, which were still sharp and unfaded.

After talking with Clara I made an effort to search out and to interview old-timers who had hunted the otters. Among them was C. B. Horn, who settled at Copalis in 1888. He had, he said, often hauled spruce poles to the beach to be used for building watchtowers fifty to sixty feet high. Working in pairs, one hunter would sit in a tower with a rifle, scanning the sea for otters. His partner would patrol the beach to retrieve any animals killed. Billy Garfield, an Indian at Taholah, let me examine one of the sea-otter rifles—a huge 50-caliber Sharps "buffalo gun" weighing 16 pounds and firing a 500-grain bullet. Charlie McIntyre built a tower on Copalis Rock, one of the many sea stacks, or monoliths, that stand like tombstones along the Olympic coast. Even today, with a good pair of binoculars, one can see the rusty iron base of his tower.

At Mukkaw Bay I talked through interpreter with Landes Kalappa, an Indian whose face stamped him as being at least seventy years old. We met him as he was returning to his driftwood shack at the edge of the bay, carrying the body of a duck he had found in the surf. He told us that, as a young man, he had often watched the sea otters feeding on "devil fish" (octopus), "sea eggs" (urchins), and crabs. Of course I couldn't understand a word he said, except once when he pointed to me and inquired, "Government-man?"

Those talks with the sea-otter hunters were an example of the para-zoological technique known as the historical interview. One of its great masters was Ernest Thompson Seton, whose four-volume *Lives of Game Animals* (1925–27) is replete with hearsay material. In employing the technique, the zoologist keeps a courteous voice while concealing critical thoughts. It is useful when he is trying to reconstruct the original population and distribution of a vanished species, such as the passenger pigeon or the wild bison. More important, it is often rewarded by a valuable anecdote from a person who has chanced to see some unusual wildlife happening. A trapper once told me that, on a certain Fourth of July, he had quietly approached a pair of mating weasels. I was skeptical, for I supposed that the weasel, like its relative the mink, must surely breed in late winter. However, I later learned that the weasel does indeed mate in summer and gives birth after a long gestation of ten months.

A Gray Wolf Skin

Soon after delving into the history of the Washington sea otters I cut the trail of another vanished animal, the Olympic gray wolf. You can locate on maps of Olympic National Park the names Cameron Creek and Gray Wolf River. In 1945 I was privileged to meet Cameron himself and to see what remained of the wolf for which the river had been named. Homesteader

STUDIES OF LAND ANIMALS

Amos B. Cameron had divorced his wife, Sarah, when both were in their sixties. She kept the old home at Carlsborg while he moved to Forks, where I interviewed him.

"It was 1919 . . . mid-November in snow," he said, "when I caught the wolf in a bear trap. He was traveling in a pack of six, following a band of elk. He was about six feet long and seemed to weigh a hundred pounds when I carried him home on my back. I had his skin mounted for a rug."

A rug? My scalp tingled. Maybe it was still stored in the old Cameron home and I would be able to see what the pelage of the Olympic wolf had looked like. I called on Sarah and found that, yes, it was there in the attic, wrapped in brown paper. I tried to buy it but Sarah wouldn't sell. (She was afraid, I think, that it might find its way back into the hands of her ex-husband.) She did let me photograph it. Warm gray in color, with a darker streak down the back, it measured seventy inches from snout to tail tip. After Sarah's death the rug went to a relative, Mrs. Hugh Cameron, who generously loaned it in 1966 to the Washington State Museum.

The Cameron wolf was among the last of the magnificent Olympic wolves. A year later, in 1920, homesteader Grant Humes trapped an eighty-six-pound male on the Elwha River and sent its skull to the National Museum. From the evidence of wolf howls, tracks in the snow, and an occasional glimpse of large, doglike animals, a few wolves may have survived into the 1930s. Then they were gone. They had no chance against the guns, traps, and poisons of civilized man.

When Olaus Murie and I were together on the *Brown Bear* he told me that, as a young man in his twenties, he had spent the winter of 1916–17 in the Elwha Valley. He was on his first assignment for the Biological Survey, namely to investigate reports of damage to elk, deer, and livestock by wolves. In those early, fumbling years of wildlife management, the *prey* of the wolf, not the wolf, was regarded as the endangered species. He saw no wolves at all that winter and grew so discouraged that he offered to resign. His resignation was not accepted.

~ There's a sequel. In 1975, eight students at Evergreen State College (Olympia) carried out a study of the feasibility of reintroducing wolves to the Olympic Peninsula. Starting with data on the potential food for wolves (thirty-six thousand deer, elk, and mountain goats, plus smaller prey species) and the potential hunting range (nearly four thousand square miles), they concluded that a starter pack of wolves might grow slowly to between forty and sixty animals, then stabilize in numbers.

The real question, of course, is whether Olympians would welcome wolves—those green-eyed, slavering predators upon livestock, pets, and babies. An Olympic old-timer whom I first met forty years ago is quoted by the Evergreen students as saying, "Anyone who'd put back wolves is a crazy!" Perhaps. I nonetheless am encouraged by the students' vision of an Olympic fauna restored to health.

An Experience Among Cannibals

The Seattle Fur Exchange is an auction house well known to fur sellers and buyers of the Pacific Northwest. In the winter of 1943–44 its manager began to complain about the poor quality of muskrat pelts that were reaching him from Tule Lake, California. About half arrived torn or bitten by some mysterious predator. Tule Lake was then open to commercial trapping under supervision of the Fish and Wildlife Service, so the Service asked me to troubleshoot.

In February 1944, while the lake was still frozen, I followed a trapper around his lines. We pushed through brown reeds along the trails of mud left by the muskrats as they came and went from their holes in the ice. We found blood spots, and the freshly torn bodies of muskrats which had been attacked in traps.

Then I climbed to the top of a muskrat lodge and looked full-circle around the horizon. I counted more than a hundred lodges extending over an open, desolate part of the lake from which most of the reeds had been removed. I knew then that I was looking at a muskrat eat-out. The population had exploded and the starving animals were turning cannibal, attacking first

STUDIES OF LAND ANIMALS

their own companions caught in traps. Although muskrats depend on plant food as a steady diet, they will, like many other rodents, become carnivores in times of stress.

When I wrote a report of my visit to Tule Lake I was unable to suggest any remedy that would satisfy all those having an interest in the lake. Clearly, more intensive trapping would bring the muskrat population into balance with the animals' food supply but it would also antagonize persons (myself included) who view commercial trapping of any kind as an inappropriate use of a national wildlife refuge. And Tule Lake is such a refuge.

In 1944 I was only beginning to understand that muskrats, like their relatives—lemmings and voles—undergo cyclical population changes which are a genetic, natural part of their "musk-

9. Tule Lake, California, 1944, where muskrats have severely cropped the cattail rushes. Three muskrat lodges are visible in the foreground.

ratishness." A bust following a boom is their way of governing their own populations. Why this should be true is unclear, though what is clear is that living processes do not have single causes. Thus, for us as humans to live comfortably with wildlife often means to accept what we can't easily or perhaps justifiably change.

4

To the Fur-seal Islands

BY the spring of 1940 I had been working three years as a federal field zoologist—years spent mainly in exploring the geographic distribution of mammals in the Pacific Northwest and the Aleutians. From study of the careers of older zoologists, I sensed that I ought soon to adopt a specialty. I thought briefly of entering the Biological Survey's food habits laboratory in Maryland, where men and women were engaged the year around in identifying the stomach contents of vertebrate animals. This was essential work and it had a sort of detective-agent appeal, but not enough to lure me to Maryland. I rejected also the thought of entering the division of predatory animal and rodent control, where my job would be perfecting methods of poisoning, gassing, and trapping nuisance animals. And I did not think seriously of becoming a wildlife refuge manager, attractive as life in the outdoors would be, for I don't like administrative work.

Meanwhile, President Franklin D. Roosevelt was making changes in the federal conservation bureaus, one of which was about to turn my career in the direction of sea-mammal research. On July 1, 1939, he had transferred both the Bureau of Biological Survey (Department of Agriculture) and the Bureau of Fisheries (Department of Commerce) to Harold Ickes's Department of the Interior. On June 30, 1940, he combined the two as the Fish and Wildlife Service.

A few older employees of the Bureau of Fisheries, especially those involved in commercial fisheries, muttered about the shotgun wedding of the bureaus. They feared that the American fishing industry was now to be under the influence of idealists who would stress the protective, as against the exploitive, side of conservation (and perhaps a Biological Survey man *would* tend to see a king salmon as an attractive form of wildlife whereas a Fisheries man would see it as money entering his net). But Roosevelt was powerful. The two bureaus remained as a unit for sixteen years and remained in the same department (Interior) for thirty years.

Ira N. Gabrielson, chief of the Biological Survey during its final years, became in 1941 director of the new Fish and Wildlife Service. Anticipating that one of his responsibilities would be to manage the Pribilof Islands (Alaska) fur-seal industry, he had looked into its history and found that no one had seriously studied the zoology of the seals for twenty-six years. In 1914 zoologists George H. Parker (Harvard University), Wilfred H. Osgood (Chicago's Field Museum of Natural History), and Edward A. Preble (Biological Survey) had been hired by the Bureau of Fisheries to develop a method of "censusing" the seal population. They had designed a rather good one; the Bureau liked it and adopted it.

By 1940, however, the census figures being compiled annually by the Bureau were beginning to look queer. They showed on paper a lively increase in the whole population yet almost *no* increase in the two classes of seals which can easily be counted—the young males, or "bachelors," tallied during the annual killings and the old breeding males, or "harem bulls." ("Harem" is the conventional name for a discrete group of female seals guarded by a breeding male.) The bulls habitually stay on land during early summer and are easily seen above the much smaller females and young.

The census figures began to look even queerer as Ward T. Bower, in charge of the fur-seal division, tried to patch up the method of computation. (He was in charge for thirty-two years, from 1915 to 1947.) Untrained in zoology, he added seals to the herd as a banker would add interest to a savings account. He supposed that the reproductive rate of the females was 100 percent, although it must have been nearer 60 percent. He placed the death rate of pups on land at 2 or 3 percent, although it was surely around 10.

Bower was extremely conservative. "I try to keep the annual

TO THE FUR-SEAL ISLANDS

reports consistent," he once told me. But, by the time the seal herd had grown on paper to 3,613,653 animals and the average bull was fathering 94.55 pups a year, zoologists were certain that the published figures were larger than life, and even Bower lost faith in his own bookkeeping.

First Trip to the Pribilofs

Gabrielson was among those who early doubted the accuracy of the seal estimates. Knowing that he was slated soon to become director of the Fish and Wildlife Service, he saw a chance to bring together, for the first time within the same agency, zoologists and administrators who could jointly take a fresh look at the seal population. He asked Frank G. Ashbrook, in charge of fur-bearer investigations for the Service, to draft a seal research plan. As a consequence, Ashbrook and I sailed from Seattle to the Pribilofs on June 10, 1940, on the *Penguin*. With us was G. Donald Gibbins, vice-president of the Fouke Fur Company, the firm that had long held the contract for processing (or curing and finishing) the sealskins.

Also on board was Harry Clifford Fassett, a gentle old fellow who appeared to be in his seventies and who had been hired as a special investigator. A link to the past, he told me that he had once served as captain's clerk on the historic steamer *Albatross* and in 1914 had been a government agent on the Pribilofs. He had written a story for the *San Francisco Chronicle*, published in 1890, entitled "Sea Otter Hunting: How the Aleuts Conduct the Chase." The date of the story, coupled with the fact that he told me of collecting plants at Unalaska in 1890, points to the certainty that he was on the *Albatross* in the Bering Sea in that year. Soon after our task force landed on the Pribilofs in 1940 he was chased by a bull seal and suffered a painful tumble to the rocks. During the rest of the summer he spent most of his time indexing, in a neat Spencerian hand, the contents of the annual *Alaska Fishery and Fur Seal Industry* reports.

~ En route to the Pribilofs from Seattle, we sat one evening on the afterdeck of the *Penguin* listening to radio news of the fall of Paris to Hitler's army. Would Britain be next? Would America now enter the war? We talked soberly as the ship throbbed along past the snowy Fairweather Range. For a while, our mission to study the fur seals seemed far away.

After we had settled on St. Paul Island, the largest of the Pribilof group, Ira Gabrielson joined us. He was in transit on an extended reconnaissance of Alaska. At a Fourth of July ceremony he saluted while the St. Paul villagers raised a makeshift and wholly unofficial "Fish and Wildlife Service" banner to christen the newborn agency. Then he addressed as Fellow Americans the three hundred mixed-race natives of the island. In the community hall that night he danced tirelessly with the native girls, thus breaking a long tradition that "we whites don't mingle socially with Aleuts." It took only one big man to break that tradition and "Gabe" was the one.

~ I dimly recall an occasion in the 1920s when my father brought Gabe to our home in Puyallup for dinner. (Both men were then working for the Biological Survey.) Gabe was a relaxed yet forceful speaker. During the meal he lambasted certain high officials in the Survey who, he claimed, were giving the agency a reputation for inertia, if not stupidity. After he left, Mother was worried. Should Father continue to be friendly with this radical, this agitator? Gabe, of course, was only expressing his concern for good wildlife conservation, a concern that was later to carry him into offices of great responsibility. In 1970 he retired from the presidency of the Wildlife Management Institute and in 1977, at the age of eighty-seven, widely honored, he died.

The Plight of the Aleuts

In the eastern Bering Sea, the volcanic Pribilof Islands stand in isolation three hundred miles from the mainland of Alaska. The largest two are St. Paul and St. George, fourteen and twelve miles long, respectively. Those two had been inhabited from nearly the time of their discovery in 1786 by the Russian fur-seeker Gerassim Pribilof. There he and his men saw

TO THE FUR-SEAL ISLANDS

what no Europeans had seen before—the summer gathering of at least a million fur seals. Although historical details are lacking, it is clear that Russian entrepreneurs (or *promyshleniki*) carried Aleut natives to the Pribilofs during the late 1700s and set them to work harvesting the pelts of the fur seals, sea otters, and foxes, as well as walrus tusks. Unpeopled at the time of their discovery, the islands now support nearly six hundred Native Americans whose family names and body features reflect, in various degrees, their Aleut, Russian, and other bloods.

When I first landed on the Pribilofs, federal control over the Native Americans, then termed "wards" though bona-fide citizens of the United States, was autocratic and discriminatory. I once heard the agent of St. Paul Island scold a native boy for a simple, friendly act—he had called me Vic instead of Doctor Scheffer! The natives were told they couldn't vote because there was no way to report the results. But why not via radio? The few white families who lived on the islands enjoyed fresh milk and imported beef, and their houses were provided with running water. The natives drank canned milk and ate fish and salted or smoked meat—meat that was mainly seal flesh that they themselves had cured during the summer.

The natives had no indoor toilets, a fact which will have sharp meaning to those who have lived through a Bering Sea winter. I can still see in memory the meat-drying racks festooned with strips of seal "jerky," the flesh black with flies, the racks only a few paces from open privies. Because strict prohibition against alcohol was enforced, the government agent weekly inspected every native house (but not the white houses) to sniff out bootleg liquor or *kvass* brewed from potatoes, rice, flour, or raisins.

The second-class status of the Pribilovians was exposed by Fredericka Martin, wife of the physician who served the islands in 1941–42. She expressed her dismay, in *American Indian* Magazine in 1946, in words that were labeled "communistic" by Washington officials. She published an article which told of feudal conditions, ration handouts by white agents, and poor schools and medical facilities.

"Is there another locality under the Stars and Stripes," she asked, "where citizens are warned that, if they insist upon leaving their birthplace, they cannot return or resume labor at the only means of earning a living?"

Stung by mounting criticism, the Secretary of the Interior appointed an eight-man team "to make a factual study of the living conditions and human problems of the natives in the Pribilof Islands and various other native communities in the Bering Sea area." The team included the director of the Fish and Wildlife Service, the Commissioner of Indian Affairs, the manager of the Pribilof Islands, and three other federal officials, as well as the director of extension of the University of Alaska and the executive secretary of the Home Missions Council of North America. No women, naturally, and no sociologists.

When I first heard how biased the team was to be I lost hope for its success. However, it did inspect the Pribilofs in 1949 and did recommend changes that were destined to give the natives greater responsibility for the handling of their own affairs. And to give them greater self-respect, which was of course what they needed more than fresh beef or ceramic plumbing.

I dwell on the old days mainly to emphasize how greatly the lot of the natives has improved. Now they vote, own title to the Pribilof soil itself (except the seal rookeries), drive cars, and own stores, theaters, and restaurants. The men and women employed at sealing receive a yearly salary in contrast to the piece-work pay they used to get. And, as recently as 1978 the federal Indian Claims Commission awarded the Pribilovians $11,239,604 compensation for damages that they and their ancestors had suffered from 1870 to 1946.

Overwhelmed by Seals

In memory, I climb again through the wet ryegrass to a cliff for my first view of Vostochni rookery. Suddenly I meet a wall of icy wind from the sea, rank with the smell of a hundred thousand bodies. I hear a great chorus, an unearthly chorus, a sound that will forever echo in my mind. I stare in disbelief at a carpet of moving forms. Black and brown and gray, they cover the beach and move among the slippery boulders and fade away a mile or more into the mist.

TO THE FUR-SEAL ISLANDS

When Ashbrook and I landed on the islands in 1940 and first saw the seals we felt crushed by the enormity of the task that lay ahead. How could we ever estimate their numbers by age and sex? They were a teeming multitude, some members of which, at any given time, were resting on land while others were coming or going between the land and the sea.

We realized that we (or the zoologists who would follow us) would need to discover certain basic facts of seal biology before the Fish and Wildlife Service could effectively manipulate the population. We asked ourselves: How could we estimate the age of an individual seal? How determine the death rate within each age class? How learn the age at which the average male or female begins to breed, and the length of its breeding life? Fur seals being polygamous, we wondered how we could estimate the optimum sex balance so that the seal clubbers could be instructed to take neither too few nor too many "surplus" males during the annual killings. (Our sole assignment was to discover facts about the seals which would support efforts to harvest as many sealskins as possible each year. There was no room for sentiment in that task.)

But in 1940 none of these data was at hand; if at hand, it would likely have been wrong. The virgin females, for example, were thought to mate for the first time at age two years. Only later did we learn that few of them do mate until they are three years old or older. The difference has important statistical implications.

A Roundup and a Branding

Ashbrook returned to Washington in mid-July 1940 and left me alone to probe the mysteries of fur-seal anatomy, behavior, and population structure. I decided to begin by counting all the pups on a sample rookery, or breeding ground. The results would at least be useful for comparison with the computed data of Bower's tables. So, on a misty morning in August, I went with a gang of natives to Zapadni Reef rookery and staged a roundup of pups. We made the little fellows run the gauntlet between two men holding tally instruments. The drive went smoothly until a

10. Fur-seal pups being rounded up for counting, St. Paul Island, Alaska, 1940.

group of pups stampeded into a tangle of boulders and driftwood. We worked frantically to dislodge them, for they were beginning to pile on top of one another in a struggling, steaming mass. We succeeded, although we learned at tragic cost a lesson in how to count pups. Forty lay lifeless on the ground. (We counted these as living individuals in the final tally.) By noon the operation was over and I had made the following comparison:

Counted pups .... 3,250 living and 196 dead
Pups according to Bower ..... 1,200 living and 10 dead

The difference was unexpected; it seemed to show that Bower's figures were *under* estimates! We later learned that the Zapadni Reef population had been growing quietly and unnoticed at the expense of nearby rookeries and was thus a poor index of the growth of the whole Pribilof population. Although the average seal is loyal to its own beach it may move away if that beach becomes crowded. Our count of dead pups did, however, indicate that the mortality rate was rising far faster in real life than on paper. And the unexpected results of the counting experiment gave us justification for continuing research.

The natives assured me that by looking at a seal they could tell its age up to five years. I was skeptical. I decided to mark a number of first-summer pups, then recapture some of them in later years and chart their age-linked characteristics. I would develop a set of imaginary, though useful, "standard seals." Accordingly, a gang of a dozen natives and I branded five thousand pups. The choking fumes rose into our faces when the iron, heated red by a torch, penetrated the wet black fur and, too often, the underlying skin. The pups bleated like lambs torn from their mothers.

In retrospect, how was I able to inflict that unintended cruelty? I was young and ambitious; scientific curiosity led me to submerge the compassion I felt. I saw the pain of the animals without admitting to myself that I was the agent of it. I had not yet read the thoughtful conclusions of W. M. S. Russell, zoologist at University College, London, in his 1959 book (with R. L. Burch) *The Principles of Humane Experimental Technique:* ". . . the greatest scientific experiments have always been the most humane and the most aesthetically attractive, conveying that sense of beauty and elegance which is the essence of science at its most successful."

After the hot-iron branding of 1940, I, or other fur-seal zoologists, searched the rookeries and killing grounds each summer for branded individuals. We found hundreds—some of which had survived for twenty years. But we hot-iron branded only one more time, in 1941. By then the Fouke Fur Company was beginning to complain that brand scars ruin the value of the pelage, while I was beginning to wonder whether a branded (traumatized) individual would mature normally and thus qualify as a "standard" seal for research purposes.

Let me jump ahead for a moment. After 1941, we turned to using a metal tag clipped to the loose skin of the seal's armpit. The tagging program grew . . . and grew . . . and grew. (Our staff biostatistician continually demanded larger and larger samples.) It peaked in the summer of 1955 with the tagging of 49,870 pups.

~ In 1940 we were still carrying on research under a handicap imposed by the 1911 Treaty for the Preservation and Protection of Fur Seals, which had blindly made no provision for killing seals for scientific study, only for commercial use. One day in late summer, as the St. Paul agent and I were walking along the edge of a rookery, we came upon a bull gravely wounded from territorial fighting. He could move only his eyes and mouth and was obviously dying.

"He's been here for two weeks," I pointed out. "I'd like to shoot him and collect his skull. We can call it a mercy killing."

"Well, go ahead," replied the agent nervously, "but don't tell anyone." Now I confess.

On another occasion, on a day when it seemed that the sun might break through the Bering Sea overcast, I drove to Vostochni rookery to photograph the harems. There I found Harry W. May, the superintendent of sealing operations, standing at the edge of the rookery with his own camera. He beckoned to me: "Come here . . . you won't believe this!" A few yards away, a bull seal was copulating with an oddly passive female. After he had dismounted, we approached her and found that she had been dead for at least a week! I should not have been surprised, for I had often seen homosexual acts by bachelor seals who were responding to inner urges they had not yet learned to interpret.

~ During my first years on the Pribilofs I planned two operations—the Lemming Caper and the Sea Lion Caper—that I now regret. They were poor zoology.

St. George Island has an indigenous population of lemmings, while St. Paul Island, forty miles to the north of it, does not. In order to furnish a new food resource for the St. Paul blue foxes, and thereby increase the yield of pelts, I arranged in 1940 to have a dozen lemmings caught alive on St. George and "seeded" on St. Paul. When the animals arrived aboard ship in their individual cages I turned them loose on a field green in tundra vegetation and richly supplied with rocky hidey-holes. They were never seen again.

What had I done wrong? In the first place, I had obtained stock from an exploding population; the St. George lemmings happened in 1940 to be in one of their cyclical peaks. According to modern theory, many individuals in such peak populations are unfit to breed; they are doomed to die from hormone stress, from a sort of internal time bomb. In the second place, I should have realized that, if St. Paul had no native lemmings, it was probably for some natural reason. Surely through the ages lemmings would have reached St. Paul from the mainland time and again on ice-rafted blocks of sod. Exactly why they did not survive and colonize is still an ecological mystery.

And I was wrong for an aesthetic reason in wanting to move lemmings from St. George to St. Paul. Had I succeeded, the two islands would now be more nearly alike. I would have destroyed a fraction of the diversity which makes Earth an interesting place.

In the Sea Lion Caper I recommended that the St. Paul natives be allowed to take a few dozen newborn pups each summer from the Northeast Point sea-lion rookery. The chocolate-brown pelts of the pups have an attractive moiré pattern. So the natives began by killing 110 pups in 1949, and, within a decade, all breeding adults had deserted the rookery. I had been instrumental in destroying the only Steller sea-lion rookery in the world which could be reached on foot by zoologists and other seal-watchers.

As I think about the Sea Lion Caper I reflect that the attitude of zoologists toward wild animals has changed remarkably in one generation. Although wild animals continue to be thought of as a "resource," they are increasingly held to be more valuable alive than dead. Zoologists of the 1940s were prone to ask: How can we use this resource? How can we harvest or crop it? Nowadays, the question is more apt to be: How can we profit from it as it is, within its ecosystem? The next question will be: Why should we expect to profit financially from it?

The sea-lion pups were killed to provide pelts, a luxury good. Had they been spared, visitors could now enjoy the seasonal drama of life on the rookery, the white gulls wheeling overhead, the foxes venturing among the sea lions in search of food, the shining kelps manured by the outwash from the beach, and all the other fascinating elements of a rookery ecosystem.

The intellectual maturing of a zoologist is a process in which his or her vision of the wholeness of life becomes sharper. It is a process in which the ancient, time-tested organic systems of Earth win more and more respect because they are seen to be complete and self-renewing. They *work*.

Fresh Eggs by the Thousands

One June morning in 1940 I was invited to take part in a bird-egging trip to Walrus Island, a small rocky plateau a few miles northeast of St. Paul Island. For many years the St. Paul natives had been accustomed to visit the island in early summer to gather fresh eggs of the common murre and thick-billed murre. The eggs were a welcome source of fresh food.

We slipped through the fog in a dory, guided to the island by its acrid smell and by overhead glimpses of murres flying swiftly to and from their nesting grounds. We landed through the surf, each person carrying an empty carton or a pail. The natives began at once to drive the murres off the nesting plateau, forcing them to huddle along the outskirts or to take wing. The birds protested in harsh, gabbling cries. Gulls swooped and screamed, trying to steal the uncovered eggs. A steady rain of gooey, fishy droppings fell among us, a hazard for which I had been warned to prepare by wearing oilskins. In about an hour the excitement was over. Some three thousand to four thousand eggs were safely stowed in the dory, most of them simply dumped in a heap in the bilge.

The annual foray to Walrus Island lives only in memory. In the 1950s the natives began to get fresh eggs from the mainland by frequent air transport. What's more important, they began to realize that the great sea-bird colonies of the Pribilofs are a national treasure, not to be disturbed lightly. This new concept reached them via the tourists and the zoologists who, in growing numbers, were beginning to visit the islands in summer.

Interludes of Beauty

I don't wish to leave the impression that I spent the entire summer of 1940 immersed in gore and filth. There were interludes of beauty. On many a day I set out in boots and windbreaker to tramp across the tundra, watching for foxes, reindeer, rosy finches, longspurs, snow buntings, golden plovers, and other animals. Waves of color rose from fields of yellow poppies and blue lupines. One flower, the alpine forget-me-not (*Eritrichium*), will always mean Alaska to me. It opens its clear blue petals a few inches above the ground in such brilliance as to give the impression of a pool of luminous paint spilled by some careless painter.

I saw relics of bygone days—the collapsed salt house at Northeast Point where the sealskins used to be cured by hand; the native dugout roofed with whalebone at Marunich; the heaps of seal bones gathered during World War I for use as fertilizer, then hastily dumped on Armistice Day; and (scuffed by my toe from the sand) a walrus-ivory doll worn smooth by the fingers of a native child long ago.

There were times when I felt utterly alone—the last living man on a planet that was beginning to cleanse itself after the departure of all other men—a planet on its way back to purity, simplicity, and goodness. Resting on some grassy hilltop, hearing only my heartbeat and the sweet voices of birds and the roll of distant surf, I felt the enormous, swelling beauty of St. Paul Island and wistfully hoped that someday it might become a national treasure valued more for its wildness than for its commercial products.

The Rise and Fall of a Reindeer Herd

While I was summering on St. Paul in 1940 I often saw at a distance a wild reindeer herd of about a thousand animals. There ten years later I counted only *eight*. What had happened? The natives blamed the decline on inbreeding, on sport hunting by soldiers during the war, on bad weather, and on disease. One morning I studied the government's reindeer records and then set out to examine the animals' rangeland, or pasture.

Back in 1911 a seed colony of four bucks and twenty-one does had been brought to the island on the revenue cutter *Bear*. (An environmental impact statement was not required at that time. More than a half-century was to elapse before the National Environmental Policy Act of 1969 would compel wildlife managers to think twice before meddling with natural ecosystems.) The deer found themselves at once in a Garden of Eden, unthreatened by predators and subject to little hunting pressure. They feasted on reindeer moss, wild parsnips, fern leaf, beach grass, crowberry, and other plants. Their numbers slowly grew to 2,046 in 1938, then began to topple. In 1950 eight deer were alive; in 1951 a single animal.

Here was a textbook example, plainly illustrated on a small treeless island, of the explosion of a pioneer population. I wondered what had tipped the balance—what had precipitated the decline of the herd. As I tramped over the tundra I noticed that reindeer moss *(Cladonia)* was now flourishing only near the village, a place the deer rarely approached. I recalled that Olaus Murie had once studied the Alaskan caribou (nearly the same as reindeer) and had discovered that *Cladonia* was the one critical food which sustained the animals during the coldest months of the year. I concluded that the St. Paul herd had destroyed itself by eliminating that nutritious and all-season food species.

The zoologist looks for reasons behind a disaster of this magnitude and tries to prevent it from recurring. The St. Paul case proved once again that wildlife management is partly people management. Had the natives been encouraged to cull the deer herd systematically, it is likely that a herd in balance with the carrying capacity of its pasture could have been maintained indefinitely. In 1951 reindeer were replanted on St. Paul and again they seemed about to eat themselves to death. When I last followed their history, they were being held in check by military riflemen.

Incredibly, in 1964 the Fish and Wildlife Service proposed to plant musk oxen on the island. Although more than a thousand miles separate the natural range of the species (on the Arctic Slope) from the Pribilofs, I suppose that musk oxen would have survived well enough on St. Paul, but, more to the point, the transplant would have been a gross insult to a unique environment. The reindeer planting had been damaging enough. The Service dropped its plan. Had it insisted on continuing, I threatened to carry a protest to the American Society of Mammalogists and to the Wildlife Society. There must have been others in the Washington Office of the Service who felt as I did, for no musk oxen were planted on the fur-seal islands.

Twenty-five years after the reindeer crash on St. Paul, zoologists learned of an even more spectacular crash on St. Matthew, another island in the Bering Sea. I began to suspect that the crash was imminent when, in the mid-1950s, I received a letter from the National Science Foundation asking me to review a request for funds from a zoologist who proposed to study the St. Matthew herd. I replied (substantially): "From the strict point of view of a zoologist it would be interesting to monitor the rise and fall of the herd along with the deterioration of its pasture. From the broader point of view of an ecologist, however, might it not be better to reduce the size of the herd immediately and thereby save the unique vegetation of St. Matthew?"

To go back a bit, the Coast Guard had planted twenty-nine reindeer on the island in 1944 as an emergency food supply for military personnel during World War II. After the war all humans departed and the deer began to multiply at a rate of about 28 percent per year. By 1963 they numbered six thousand. The next winter was a terribly cold one, and with the arrival of spring, the deer had been virtually wiped out. Fewer than a hundred remained. (Strangely, all but one of the survivors were females, while the lone male was evidently impotent; it had abnormal antlers.) In 1977 two zoologists who were counting sea birds on the island saw only nine old deer—tragic remnants of the St. Matthew herd.

Threatened Collapse of the Fur-seal Treaty

Conservationists supposed that the four-party fur-seal treaty of 1911 would be renewed in 1941 as it had been in 1926. They saw it as a necessary agreement to protect an international resource, a common heritage. Unexpectedly, on October 23, 1940, Japan gave notice of her intent to withdraw from the treaty one year later. That action would automatically bring about its collapse.

The treaty had been designed as an elaborate tradeoff in which the "have" nations (United States and Russia, owning fur-seal grounds) had promised to pay the "have nots" (Canada and Japan, lacking such grounds) for agreeing to end pelagic sealing. (This statement is not *quite* true, for Japan had seized tiny Seal Island, off Sakhalin, during the Russo-Japanese War of 1904–1905 and the island was the summer home of a few hundred seals.) Pelagic, or oceanic, hunting of seals is wasteful, for many of the wounded animals sink before they can be recovered, and it is nonselective, for many of its victims are pregnant females. These are more valuable as breeders than as a source of pelts.

The treaty was coldly commercial—devoted solely to protecting a resource worth millions of dollars. It made no mention of aesthetic, educational, or scientific values of seals. (It did allow sea-coastal "aborigines" to continue hunting them for subsistence purposes.) It did not touch on the importance of

11. A fur-seal pup (*right*) is being born on St. Paul Island, Alaska, 1965. It weighs 11 to 12 pounds; its mother weighs about 100 pounds.

seals as active elements in a vast, rich, recycling marine ecosystem.

Japan went along with the treaty, although with less enthusiasm than any of the other three signatory powers. As an island nation she found it painful to quit pelagic sealing and thus to compromise the ancient principle of freedom of the seas.

So by 1940 she had decided to pull out of the treaty. She claimed that the American seal herd, then grown to an estimated 2,200,000 animals, was eating thousands of tons of valuable seafood which otherwise might have found its way into Japanese nets and onto Japanese tables. Ironically, the inflated census data that Ward T. Bower so proudly published each year served, in the end, to undermine the bargaining strength of the United States at the treaty table.

Japan's grievance, moreover, was based on her conviction that Pribilof-born seals habitually wander westward to feed in Asian waters. The United States denied it. Our official position was that the three herds of seals known to breed on Japanese, Soviet, and Alaskan islands, respectively, belong to three distinct races which do not fraternize at sea. That position had to be abandoned shortly after World War II, when it was learned that Eastern and Western seals do intermingle extensively and that the birthplace of a fur seal cannot be identified from its body appearance or structure.

But surely our State Department, long before 1940, must have known that the seals intermingle, for branded animals of American origin had more than once been reported from Asian shores and waters. They were some of the nearly 35,000 individuals branded for one reason or another between 1896 and 1929. Being a minor employee and living far from the city of Washington, I was not privy to State's information. I learned about intermingling by writing directly to the Pacific Institute of Fisheries and Oceanology (TINRO) in Vladivostok, asking, "Please, have you seen any of our branded or tagged seals?" A courteous reply came in 1945, giving details of the marks which Soviet biologists had seen on the Commander Islands. (The year 1945 was one of unusually cordial relations between East and West.)

For failing to route my correspondence through the State Department I was rebuked by a liaison officer in Washington. He used the words "close to treason." Be that as it may, writing to TINRO seemed like a good idea at the time, and it brought results.

5

Locked Onto Sea Mammals

IN the winter of 1940–41, staff members of the Fish and Wildlife Service in Washington met to plan a crash program of research on the Pribilof seals. The results, they hoped, would dissuade Japan from pulling out of the treaty. (They did not see the war clouds forming.) As they planned it, the research would measure the damage caused by seals to commercial fisheries and would measure the degree of intermingling of Asian and American herds. Further to placate Japan, the United States would reduce the size of the Pribilof herd.

In December 1940, Hartley Jackson wrote to me from his office in the National Museum:

> For many years the study of marine mammals has been seriously neglected by American mammalogists. There has in recent years been only one man, namely Dr. Remington Kellogg, who has made any serious study of the cetaceans, and his work has been confined largely to taxonomic and paleontological studies, in which we are all willing to admit he has done a splendid job. Pinnipeds have been almost entirely neglected since the days of Dr. Frederick W. True. With the unity now existing between the former Bureau of Fisheries and Biological Survey . . . it seems not only desirable but essential that we have someone in our organization qualified to give information on marine mammals and to conduct any administrative studies or general research that may be necessary. . . . How would it appeal to you to devote your efforts primarily to this line of work? It would undoubtedly be necessary for you to move to Washington, since all the facilities for such studies, particularly as to libraries and specimens, are here."

LOCKED ONTO SEA MAMMALS

So the die was cast. I was to specialize in studies of the beasts of the sea. Looking forward with pleasure to working in the National Museum among the great zoologists and the ghosts of their predecessors, I sold a newly built house in Seattle and prepared to move to Washington. In February 1941 I went there to talk with Jackson, Kellogg, and others about specific research plans.

Scientists in the City of Washington

Kellogg was the fourth in line of the zoologists who had assembled the mammal collections of the National Museum; he had been preceded by Spencer Fullerton Baird, Frederick William True, and Gerrit S. Miller. He was a big man—kindly, persuasive, and fluent in the milder dialects of profanity. He was also enormously productive. Associated with the museum for nearly fifty years, he was its director from 1948 to 1962. His doctoral thesis, "The History of Whales —Their Adaptation to Life in the Water" (1928), is still the best summary of that subject. The publication in 1936 of his monograph "A Review of the Archaeoceti" established his reputation as the world's leading student of whale evolution.

In April 1930 he went to Berlin to the first international (League of Nations) conference on whales and whaling, a trip which was to lead him deeper and deeper into whaling politics. He was a forceful member of the International Whaling Commission during its first eighteen years, from 1949 through 1966, when he resigned at age seventy-three. He was succeeded by J. L. McHugh (1967–71), by Robert M. White (1972–76), by William Aron (1977), and by Richard A. Frank (1978–   ).

The first time I talked with Kellogg he hinted that among the irritants in his busy life were people who sent him manuscripts to be reviewed and others who sent him bones to be identified. Having warned me, he cheerfully performed both services for me again and again.

ADVENTURES OF A ZOOLOGIST

~ While I was in Kellogg's office one day a stately, white-haired gentleman drifted in for a chat. He was Leonhard Stejneger (1851-1943), the man who had written accounts between 1884 and 1925 of the Russian seal islands and of the giant sea cow *(Hydrodamalis)*, a beast exterminated long before its time. I was greatly impressed, for I had used his 1936 book, *Georg Wilhelm Steller: The Pioneer of Alaskan Natural History*, as a reference during the Aleutian Expedition.

I inquired about his adventures as a young man in the Bering and Okhotsk seas. He told, with twinkling eyes, of hijacking sea-cow bones from Russian workmen who had dug them from the sands of Bering Island and had boxed them for shipment to St. Petersburg. Through the quiet exchange of a handful of gold, the bones were redirected to Washington to become a composite skeleton which can be seen today in the National Museum. Official reports do not mention the skeleton deal, which was probably effected in 1882 or 1883.

~ It was also my privilege to meet in Washington Ernest P. Walker, assistant director for twenty-six years of the National Zoological Park. A slight, balding, intense man with a ready smile, he and I found at once that we shared an interest in animal behavior. He was a student of living animals, not their remains in museum cabinets. He insisted that museum people know too little about real animals. Writing to me later for information on Alaskan sea life, he hoped that I might have picked up certain facts from Native Americans which, in his words, the "derm/osteo/ologists [skin-and-bone specialists] have not learned." He had gained his encyclopedic knowledge of mammals partly via the academic route but mainly through study of hundreds of species alive in the wilds of Alaska and in the National Zoo.

And also in his own home; his apartment was a mini-zoo. Once when I called there he released a tame bat that flew back and forth the length of the living room, swooping now and then to snatch a mealworm from Ernest's fingers. On another occasion he placed in my hand a tiny half-ounce kangaroo mouse, which I clumsily let fall to the floor. As it lay unconscious Ernest gently palpated its chest until it revived. While I was in the bathroom, a flying squirrel with pale, silky fur and luminous eyes startled me by landing soundlessly upon my shoulder.

Ernest was especially fond of an owl-faced night monkey, or douroucouli, that entered his childless life and soon became a member of the family, even to the point of using the family toilet. I once attended a scientific meeting in Colorado at which Ernest gave a paper, illustrated with sound tapes and visual sonograms (voice prints), describing his studies of monkey language.

"I shall offer," he began, "a Douroucouli–English dictionary."

While still at the meeting, he telephoned words of comfort, both human and simian, to his pet in Washington.

Because Ernest was a loving person, he was sure that his affection was reciprocated by his animal friends. He identified nearly fifty vocal sounds produced by the monkey. When it snuggled close and said "O-o-ohh, o-o-ohh," it was saying, declared Ernest, "I am so happy" or "I love you greatly."

Because of his tendency to anthropomorphize, however, some of his colleagues thought that he was skating rather close to the thin ice of unprofessional zoology. But in 1964 he proved his worth: Johns Hopkins Press published the magnum opus on which he had been quietly working for thirty years—his three-volume *Mammals of the World*. Fairfield Osborn, president of the New York Zoological Society, sponsor of the work, called it "heroic . . . a quite extraordinary achievement." It was soon widely referred to by specialists and lay readers alike and in 1975 was republished in a third edition.

When I last saw Ernest in 1968 we dined at a Maryland restaurant. With a quirky smile he pulled from his pocket the outline for a new book to be called *Be Kind to the Earth*. It would appeal to conservationists everywhere, urging them to work toward the preservation of nature for ethical, as well as practical, reasons. I am sure that Ernest foresaw the essentially religious core of the ecological movement, which was soon to be celebrated on Earth Day, April 22, 1970. I am sure that he believed, along with René Dubos, biologist at Rockefeller University, that evolution has created our emotional powers as well as our opposable thumbs and our cunning for survival. When, in 1969, Dubos gave a talk entitled "A Theology of the Earth" he used

12. Ernest P. Walker, assistant director of the National Zoo, in Washington, photographs a mink, 1941.

without embarrassment the words "sacred" and "worship" to describe his attitude toward our planet.

Had he lived, Ernest would have been tickled by today's revival of interest in the reasoning powers of nonhuman animals. Donald R. Griffin, animal behaviorist at Rockefeller University, is among those who believe that the higher beasts *are* self-aware and *are* able to think conceptually. "As theories are developed and tested to account for the behavior of animals," wrote Griffin in *BioScience* Magazine in 1977, "it may turn out to be more parsimonious and considerably more fruitful to consider the possibility that evolutionary continuity applies not only to [body] structure and function but also to cognition."

Training to Specialize

After talking with Washington zoologists in the spring of 1941 I resolved to cram on sea mammals, so in May I drove to the whaling station of the San Francisco Sea Products Company at Humboldt Bay, California. The director of that company believed that a diesel-powered fishing vessel could be used as a whale catcher while its crew would ordinarily be idle outside the fishing season. Steam-powered catchers were at that time universally used. He predicted (rightly) that a diesel craft, although noisier, would not be any more alarming to whales. He let me travel for two days on the trial run of the 77-foot *Hawk II,* first exacting my promise that I would never divulge the secrets of the *Hawk*'s experimental gear. The gunner shot at a sperm whale but missed it when it crossed under the ship's bow. Many years later I wrote, in *The Year of the Whale,* a semifictional account of that—my one and only—whaling voyage.

Also in the spring of 1941 I visited Neah Bay, Washington, where the Makah Indians had for years untold launched their cedar canoes and gone out to hunt fur seals on the treacherous Pacific. From 1920 to 1926, while post–World War I fur prices were high, the Olympic Coast Indians speared and recovered over a thousand seals a year. I watched a sealing party return to quiet Neah Bay at dusk on the fifth of June 1941. In their catch was the body of a pregnant female. The crew gave me her fetus, a black pup that would have been born on the Pribilofs a month later and a thousand miles farther north.

The Makahs did not seal in 1942, for by then the Olympic Coast was a tight military zone, nor did they resume sealing after the war. I was lucky to have seen the sealing ritual—for that's what it had become—before it disappeared.

Like the Makahs, the neighboring Quilieutes carried on pelagic sealing under treaty rights. (On one occasion when I

13. A gunner demonstrates the aiming of a harpoon cannon on the whalecatcher *Hawk II,* Fields Landing, California, 1941.

stopped at the village of La Push, the members of the tribe, young and old, were seated in a circle, feasting on boiled seal meat and canned beans!) Carl E. Gustafson, anthropologist at Washington State University, made a puzzling find near here at the site of a village which had been abandoned in the early 1900s. He found fur-seal bones and teeth marking two thousand years of prehistoric and historic sealing. About half the male skulls were those of adults, which is strange, because nowadays few bulls migrate south of the Gulf of Alaska. One wonders, Did the bulls change their migratory route in response to some change in ocean temperature? Less likely, Did persistent sealing genetically "educate" the oldest males, gradually eliminating a strain whose members resorted to warmer waters in winter?

LOCKED ONTO SEA MAMMALS

In June 1941 I surveyed the jagged rocks fronting the Olympic seashore between the Devil's Graveyard and La Push, hoping to learn whether the sea lions known to rest there were breeding animals. (If so, June would logically be their pupping season.) On a bright, calm day, a Coast Guard cutter took me from Port Angeles to Astoria. During part of the trip, while the cutter ran safely offshore, her motor launch carried me on a parallel course among the rocks and kelp. I saw no signs of breeding animals, nor has anyone since; the sea lions of Washington are evidently seasonal stragglers from British Columbia or Oregon.

I repeated the survey in March 1944, this time in a Coast Guard airplane with Preston P. Macy, superintendent of Olympic National Park. Our pilot swooped and spiraled to give us a close look at the rocks. When I wasn't face-down in a nausea bag I was counting sea lions. Aerial reconnaissance has its good points and its bad points.

Planning the Black Douglas Expeditions

Meanwhile, in the city of Washington, a strategy for preventing the collapse of the fur-seal treaty had finally been mapped. On March 7, 1941, Secretary of the Interior Harold Ickes wrote to the Secretary of State and to the President, outlining that strategy. First, the upcoming (1941) kill of seals would be increased by thirty thousand "in the direction of reducing the abundance of seal life." Second, a two-year investigation would be launched. These two moves were aimed at placating Japan. Cutting down the size of the Pribilof herd would in theory reduce the damage done by seals to Japanese commercial fisheries. The investigation would provide firm evidence (from the recovery of tagged seals) of the degree of intermingling between Asian and American stocks. It would also disclose exactly what species of seafoods the seals do eat, as well as when and where they take them.

At an outlay of $394,055 during the first year, the Fish and Wildlife Service would undertake to buy a research vessel of the tuna-clipper type, about 130 to 150 feet in length, at a cost of

ADVENTURES OF A ZOOLOGIST

$300,000; spend $58,845 for operating it; and hire a staff of five zoologists, with assistants, at a cost of $35,210 for salaries and expenses.

On June 30, 1941, Congress appropriated $290,000 for the investigation. Although I wasn't sure that I was ready to carry the responsibility, I was placed in charge. The headquarters of the investigation would be in Seattle, a seaport city growing rapidly in importance as a fisheries-and-oceanography research center.

So, after all, I did not move to the National Museum in the city of Washington.

Of the five new zoologists we hired, three were University of Washington graduates: Wilbert M. Chapman, A. Henry Banner, and Kelshaw Bonham. Two were from the East: Ford Wilke

LOCKED ONTO SEA MAMMALS

and Donald D. Shipley. Our headquarters was the University of Washington's Anderson Hall.

By a turn of fate, our research vessel was not to be a tuna clipper but a luxury motor schooner, the *Black Douglas,* built as a private yacht in 1930 by the Roebling family. (John A. Roebling's Sons built the Brooklyn Bridge.) She was rated at 360 tons and carried ten sails on three masts. Her decks were teak, while down below she boasted an electric piano and two bathtubs! Wib Chapman and I first saw the *Black Douglas* in early December 1941 when she reached San Diego after coming from dry dock in Savannah, Georgia. But her beauty was to be short-lived. When we finally were able to use her as a floating research base she had been drastically changed.

Japan struck at Pearl Harbor on December 7 and, a few days

14. Indians at La Push, Washington, make a seagoing dugout canoe for fur-sealing, 1930. (Photograph: Theophilus Scheffer)

later, Wib and I rode the *Black Douglas* to Seattle. It was my first voyage on a sailing ship. We were so anxious to get home, however, that we ran under full engine power with only a staysail to hold us against the December westerlies. All the shore lights were blacked out and the radio beacons were stilled; the captain anxiously watched his fathometer. Bundled to the eyes in rain gear, I took my turn at the wheel, watching the sail and the sliding stars, worrying about Japanese submarines . . . and wondering what might be salvaged from the expedition that had begun so brightly.

In Seattle we turned the *Black Douglas* over to the Navy for wartime use as a PYC (patrol yacht, coastal). She came back to us, raped, in 1946. She now carried an ugly house and her masts were gone; her warm brown teak was painted over; she had become a workboat.

~ During the summer and fall of 1941 Wilke and Banner had gone up to the Pribilofs to carry on research among the seals during their breeding season. When war broke out they returned to Seattle as soon as they could, arriving on a military ship on January 3, 1942. They escorted on deck seven live seals destined for the Seattle Zoo. We proposed to keep track of the food eaten by these so that we could roughly estimate the amount eaten by the whole seal herd in the course of a year.

But none of the captives lived as long as a year; all were dead by the end of sixteen weeks. After living two months in captivity, one voided a volcanic pebble—a souvenir which it must have picked up on the Pribilofs. Another fed normally for seven weeks, during which time it ate five to seven pounds of fish daily, representing one-seventh of its body weight.

That feeding experiment was repeated in the Seattle Zoo and in other zoos, with the result that we are now able to estimate that the Pribilof seals eat at least one million tons of fish and squid a year. They may eat twice that much; we still know little about their food requirements during winter while they are actively chasing their prey at sea and while their bodies are continuously losing heat to the chilly water.

When we first put the seals in the Seattle Zoo we posted a sign—EXPERIMENTAL ANIMALS: PLEASE DON'T FEED—which brought a note from my Washington office reading, approximately, "Scheffer, what are you doing to those animals? We have a letter here from the [Humane] Society complaining that you're using them in experiments." I quickly reported the facts. Although I sympathized with the aims of the Society I felt that its spokesperson had, in this instance, taken a devious way of getting at the truth.

Research in Wartime

In early 1942 we learned that the Pribilofs were to be occupied by American troops and that all the inhabitants were to be removed. Indeed, on June 16, the 498 men, women, and children living on the Pribilofs were loaded on a ship bound for Funter Bay, southeastern Alaska. For a week in June, caretaker Roy D. Hurd, left behind, was the only human on the Pribilofs. The 1942 sealing season had already opened but was abruptly closed with 150 skins in the salt bins.

We six zoologists of the *Black Douglas* group, marking time in Seattle, busied ourselves at miscellaneous research tasks. Wib Chapman dissected and sketched the bony skeletons of various fishes known to be eaten by fur seals. He knew that eventually we would be faced with the problem of identifying food fragments in the stomachs of seals collected at sea. He presoaked each fish in a caustic solution containing a red dye that stained the bones, but not the flesh. The specimens came out beautifully translucent, rather like X-ray images.

Wib discovered that the mysterious "seal fish," known only from specimens collected in the 1890s from fur-seal stomachs, was actually a well-known deep-sea form, *Bathylagus*. We wondered, Do the seals feed in deeper waters than anyone had supposed? That question was not to be answered until 1975, when zoologists on the Pribilofs recovered a depth gauge from a seal's neck and learned that the animal had been feeding 190 meters (623 feet) below the surface of the Pacific Ocean. An astonishing find.

After Wib left our group in 1942 he took on more demanding responsibilities in marine fishery conservation, becoming dean of the college of fisheries at the University of Washington and, later, fisheries adviser to the Secretary of State. In State, he argued his points with clear scientific reasoning and in direct—even blunt—language, thereby winning the adoption of many international agreements for the protection of North Atlantic and North Pacific fisheries. After his sudden death in 1960 one of his friends wrote that "the Department of State never recovered from him." Unused to plain language, the Department was shaken by it.

During the war, John W. Slipp, a fisheries student at the University of Washington, was a frequent visitor at our office. He showed such a keen interest in our work that we arranged to have him appointed as a Fish and Wildlife Service collaborator. He and I began to assemble information on the marine mammals of Washington State, and eventually published our findings in two technical articles, one on whales and dolphins, and one on harbor seals.

John was good at tracking down historical data in old newspapers and magazines and at prying anecdotes from residents of the seacoast. One of his rewards was a photograph of a rare whale, *Pseudorca*, which had been shot near Olympia by trigger-happy citizens. That photo, and a few moldering bones recovered from a chicken yard, still represent the only record of the false killer whale north of southern California.

On the beach at Grayland in the early summer of 1942 we found the body of a female harbor porpoise with a shad lodged in her throat. She had tried to swallow the coarse-scaled fish and it had choked her. Exactly ten years later, at the same season, on the same beach, we found another female harbor porpoise choked by a shad. A coincidence—or was it only that? I would prefer to believe that on each occasion the mammal and the fish, following convergent migratory tracks through the sea, had arrived by zoological predestination at a rendezvous with death.

~ I am especially fond of harbor seals. Long under attack by fishermen and target shooters, these shy spotted animals of the northern hemisphere still survive today more than a half-million strong. Like the wily coyotes of America's western plains, they have accommodated to life at the fringe of man's civilized world.

15. Terrell C. Newby prepares to tag a newborn harbor seal on a Puget Sound island, 1970.

Let me jump ahead to 1970 and a night spent in a seal-watching tower on Gertrude Island, near Puget Sound's McNeil Island federal penitentiary. The tower had recently been built on a sandspit by the government, in cooperation with the University of Puget Sound, to aid seal watchers with their research. Because boaters (and especially boaters with guns) are forbidden to approach the penitentiary island, a colony of several hundred seals has flourished here for many years. Terrell C. Newby, then a student working for his master's degree, spent the night with me in the tower.

ADVENTURES OF A ZOOLOGIST

Throughout the darkness, little sounds rose to our ears—the lapping of waves, the night calls of white-crowned sparrows in the forest behind us, the comfortable grunts of seals wriggling into their sandy beds, and now and then the plaintive "kro-o-oh" of a newborn pup searching for its mother's nipples. Several times, Terry crawled from his sleeping bag, peered through a window, and noted the shifting of the seals with the movements of the tide.

Harbor seals in the wild are difficult to approach. Here in the tower I was able to see individuals from a distance of less than fifty feet. When daylight came, I witnessed a birth. It was a rare privilege, for the birth of most harbor seals is remote and private, announced only by the gulls or crows that wait to seize the afterbirth.

~ In June 1942, only a day before the end of the federal fiscal year, we learned that a wartime Congress had lost interest in seal research and was not about to renew funding for the *Black Douglas* studies. Some of us scrambled for new jobs while others went into military service. Because I was thirty-five years old and had a wife and by now two children, I was not drafted. And because I was a permanent employee, the Fish and Wildlife Service regarded me as a "musty" (must-be-employed-if-possible). The Service transferred me to another branch—the division of fishery industries, where I was assigned the job of studying seaweeds useful to the war effort. Quite a change from zoology.

It seems that the outbreak of war had brought a sudden halt to the importation of a critical marine product known as agar, a colorless gum resembling gelatin. It is extracted from red seaweeds of the genus *Gelidium* and is widely used in public health and in dental laboratories. Before the war, about 92 percent of the agar used in the United States came from Japan and China; the rest was manufactured in southern California from weed gathered incidentally by abalone divers. In July 1943 I moved with my family to La Jolla, California, to an office-laboratory provided by Scripps Institution of Oceanography.

My job was to stimulate the production of agar and to find hitherto unused seaweeds, rich in agar, which would be cheaper than *Gelidium* to gather. The latter had to be picked by divers

LOCKED ONTO SEA MAMMALS

wearing deep-sea suits, working from air-compressor boats. Because my assignment in La Jolla lasted only six months and had nothing to do with zoology, I shan't describe it. While living in southern California, however, I met three scientists who impressed me deeply.

Scientists in California

One day I visited the marine biological station of the California Institute of Technology at Corona del Mar to inquire whether anyone on its staff was interested in seaweed research. Although no one was, the visit paid off, for old Thomas Hunt Morgan (1866–1945), Nobel laureate and pioneer of genetics research, was standing there quietly among the aquariums. We shook hands, no more: his thoughts were elsewhere. He was a figure who had stepped out of the pages of my zoology textbooks, been given form and substance, and was destined in a moment to return to similar pages, where, among his fruit flies and white mice, he will live for ages.

Morgan's important studies, carried out at Columbia University between 1910 and 1920, had cleared up many of the mysteries of inheritance. He proved what is now common knowledge—that the characteristics of organisms are passed on from parent to offspring through chromosomes, each of which contains genes, or hereditary units, in linear array.

I recently read an amusing story about Morgan—one that tickled me because it conjured up my lantern-slide days at Mount Rainier. His biographers, Ian Shine and Sylvia Wrobel, wrote in *Thomas Hunt Morgan, Pioneer of Genetics:*

> About 1916 Morgan . . . illustrated a talk by projecting *Drosophila* on the screen. At the moment before showing, the etherized fruit flies were inserted into a double slide with an air space. The slides were then sealed. As the flies awakened the slide was projected, and huge, live *Drosophila* with their red eyes showing crawled briskly back and forth across the screen. The effect was startling

and the audience loved it. The next slide was shown without delay for the flies were killed by the projector's heat within ten seconds after their performance.

~ At La Jolla one evening beneath the palm trees, our Scheffer family enjoyed dinner with the director of Scripps Institution, Harald U. Sverdrup, and his charming wife, Gudrun. During dinner, Sverdrup told us in his soft, sure voice something of his past. He had been a member of the famous North Pole Expedition of the *Maud,* led by Roald Amundsen between 1918 and 1925. Seven years in isolation with the same companions! In 1931 he had sailed with Sir Hubert Wilkins on the ill-fated attempt of the submarine *Nautilus* to reach the North Pole under ice.

In the early forties at Scripps he wrote the leading chapters of a textbook, *The Oceans.* How I wish that I'd bought a copy and had it autographed! Then it seemed too costly for a young zoologist supporting a family on $3,200 a year. In 1948 one issue of the *Journal of Marine Research* was devoted to an appreciative review of Sverdrup and his work. His long years on the ice, concluded one reviewer, had developed to a remarkable degree the man's "internal harmony and simplicity." Delightfully put.

~ Another friend we made at Scripps was Carl L. Hubbs (1894–1979), who was in the process of transferring from the University of Michigan. He was widely known as an ichthyologist. Starting as a student in 1915, he began to collect freshwater fishes from the continents of the world in order to map their distribution. To determine the family relationships of certain species, he bred them in the laboratory, thus learning which ones could interbreed and which ones could not. During one of those experiments he discovered that the Amazon molly *(Poecilia formosa)* is an all-female species. It reproduces by parthenogenesis. He also studied the pupfishes, those tiny fishes that live as solitary relics of the Pleistocene in cave pools and desert springs of the American West. Much to the annoyance of land developers who would like to tap the water stored in pupfish pools, the fishes are protected as an endangered species. As a result, one

LOCKED ONTO SEA MAMMALS

reads such stormy headlines as: TWO-INCH FISH BLOCKS PATH OF PROGRESS.

After Carl moved to La Jolla in 1944 he quickly added to his many interests the study of Pacific Coast reptiles, birds, mammals, and prehistoric Indians. One of his adventures in zoology involved a small whale which, when it was first spotted by beachcombers, was floundering in the surf near Scripps Institution. Seven men succeeded in roping it and hauling it to the beach, where it soon died. Carl saw that it was an uncommon species of beaked whale *(Mesoplodon)*, so he hastily rounded up a butchering crew to salvage its skeleton for the National Museum. Later the crew rendered three gallons of clear yellow oil from its blubber.

"The meat," wrote Carl in a scientific report, "was very red and turned blackish on holding, but was of good flavor and tender when roasted or fried. About 100 pounds were eaten by local residents. This addition to the war-rationed meat supply was much enjoyed."

The salvage crew also sold the oil to the war-ration board in exchange for "red stamps"—those precious coupons required for the purchase of butter, meat, or lard. All patriotic housewives were then saving bacon grease, rancid butter, and other surplus fats for use in the war effort—to be converted, I believe, into nitroglycerin.

6

Back to the Seal Islands

THE WAR was still raging when I returned to the Pribilofs in the summer of 1944 on the second of fifteen visits to those foggy isles. By then, however, the fighting had moved south of the Bering Sea. The Pribilofs were no longer under military occupation.

My research objectives had not changed. They were, first, to census the seal herd and, second, to obtain biological data on growth rates, age at sexual maturity, length of life, sex ratio, causes of mortality, territorial behavior, daily and seasonal movements of the animals, and other desiderata on a long "shopping list." All were to be synthesized as a means of obtaining the *maximum sustainable yield* (a fisheries term) of sealskins of commercial quality.

Working alone, I set up a laboratory in a dark, unfinished basement room under the St. Paul Village store. It had no running water; a rainwater barrel under the eaves served as a laundry tub for bloody towels. It was warmed by a stove which glowed redly when I fed it strips of seal blubber. Fine, black volcanic sand blew steadily through cracks in the walls. The following year, zoologist Norman O. Levardsen was assigned to help me. The seal research program continued to grow; today it utilizes the services of a dozen full-time men and women and operates on an annual appropriation of over a million dollars.

Now, it is one thing for a zoologist to work unassisted with laboratory mice or guinea pigs but quite another to struggle with marine mammals, for these are heavy, bloody, and greasy. In the spring of 1944, before the seals had returned to the islands from their winter migration, I visited a sea-lion rookery

BACK TO THE SEAL ISLANDS

at Northeast Point to learn about pupping dates and newborn sizes in that species. I found a newborn pup among the rocks and decided to take it alive to the lab for study. Because I had left my truck behind, where the road was blocked by snow, I had to carry the pup, bawling and squirming and upchucking sour milk, on my back for several miles. It weighed forty-nine pounds.

The Pribilofs were a splendid platform of opportunity for studying wildlife zoology. Where else but on those islands could a zoologist observe at one time and place the behavior of tens of thousands of mammals? Where else could he or she have daily access to hundreds or even thousands of fresh anatomical specimens? Yes, there were days during the sealing season when the bachelors came in wetly from the sea in unexpected numbers, to be met by the native gangs and soon be reduced to skins and white carcasses arranged in rows on the killing fields. In July 1949 the natives skinned a record 5,329 seals in one day.

Every summer we would see on the rookeries several abnormal seals of a type known locally as "big cows." The zoological nature of those freaks was still a mystery, so we set out one morning to collect a specimen for examination. A big cow is an awkward, shambling creature with flippers that seem too large for its body. It lacks the mane, or crest, of stiff gray hairs on the back of the neck. In the native lore of the day a big cow was thought to be the hybrid offspring of a bull seal and a female sea lion.

Well, we saw one being guarded by a bull; she was the only animal in his harem. We shot her and carried her off the rookery on a stretcher; she weighed 222 pounds. Upon autopsy, however, "she" proved to be a male with infantile testes, that is, a cryptorchid. We deduced that its body odor, being nonmale, had been translated as female by the bull who guarded it.

Although the fur seal had been known to science since Georg Wilhelm Steller first gave an account of the species to the Imperial Academy of Science in St. Petersburg, Russia, in 1751, and countless thousands of its kind had been handled by hunters, no scientist had ever described its early fetal stages. Thus it was with keen interest that we autopsied a pregnant female

16. Author in the fur-seal laboratory, St. Paul Island, Alaska, 1945.

taken from a shark net off the Oregon coast in the winter of 1945. She was carrying a ten-inch fetus, still in its first trimester yet already having well-formed ears, claws, and whiskers. It was otherwise hairless, and its eyes had not opened. Its unfortunate mother had become tangled in a bottom net at 35 fathoms (210 feet).

She was an indirect casualty of the war. To meet a national demand for vitamin A, fishermen along the Pacific Coast had begun to net soupfin sharks and to sell their livers, rich in that vitamin, for prices up to nine dollars a pound. In the late 1940s, however, chemists found a way of synthesizing vitamin A cheaply and the demand for its natural sources plummeted.

In the meantime, we collected and froze four hundred pounds of seal livers from the Pribilofs and analyzed them in

BACK TO THE SEAL ISLANDS

Seattle. Their vitamin A content ranged widely—in the order of 1 to 1,000. To this day I have not found a biological explanation for the wide variation in vitamin A content of seal liver. I suppose there may be body reservoirs in which the vitamin is stored between feedings, with the consequence that a fasting individual steadily depletes its stores.

~ On August 6, 1945, I was drinking coffee in the home of a radio ham on St. Paul Island, waiting for the evening newscast from Anchorage. Harry Truman came on the air to tell the world of the first atomic bomb ever used against human life, at Hiroshima. My first reaction was one of relief; the war would soon be over. Not until later did I understand that a black and awful entry had been made in the pages of American history—the history of a nation that calls itself a leader among civilized peoples.

17. Author photographs thick-billed murres on St. Paul Island, Alaska, 1947. (Photograph: Karl W. Kenyon)

ADVENTURES OF A ZOOLOGIST

The Health of the Seals

From the start of our seal research our main goal had been to understand population size within age and sex classes. To reach that goal we knew that we would have to study reproduction (which adds to the population) and mortality (which subtracts). When we first gave thought to mortality we supposed that the seals are remarkably healthy. They live shoulder-to-shoulder on frightfully smelly, filthy grounds, yet have never been known to suffer an epizootic disease. But, as time went on, we learned that they do in fact harbor many kinds of pathogens and parasites. About 70 percent of all those born die before reaching their third birthday—most of them at sea, their passing unseen by man.

We singled out the hookworm disease of seals for special study. The needle-sharp worms drill through the animal's gut and bring death from hemorrhage. Although the worm had been discovered in 1897, little attention was given it until 1944, when I rediscovered it by accident. (I stepped on a carcass.) I hastily picked up ten carcasses at random and found that all were heavily infested. That discovery aroused excitement in the Washington office, along with speculation that a parasitic scourge might be in the making. Authority was eventually granted to hire a hookworm specialist. So, in 1951, O. Wilford Olsen, a zoologist from Colorado Agricultural and Mechanical College, came to the Pribilofs to study the life cycle of the worm and to find a means of checking its ravages. Little did he know that ten summers would pass before he would solve the riddle of its life.

He soon found that the worms—unlike other hookworm species—can survive subarctic winters as larvae in frozen ground. He tried spraying the rookeries with various pesticides but none reduced the mortality; in fact, the count of dead pups in 1956 reached an all-time high of 120,000! Olsen was deeply discouraged.

Then, in the summer of 1961, with the help of a doctoral candidate, Eugene T. Lyons, Olsen discovered that millions of larval worms live in a dormant stage within the blubber of nearly all adult seals. This means that, when a mother nurses her pup,

BACK TO THE SEAL ISLANDS

larvae swarm out of the mammary blubber into the milk and thence into the intestines of the pup, where they attach and begin to suck blood. Those larvae that overwinter on the rookeries are not, after all, important; killing them with chemicals could never break the life cycle of the parasite.

Hookworm disease, it would seem, is a fact of life that seals, as well as we humans who are trying to care for the seals, must accept.

Nearly every summer for ten years, starting in 1944, we saw seals wearing mysterious rubber collars. The collars were alike, each being a soft, brown rubber ring like the rolled top of a stocking. Often the collar would have worked its way through the fur and into the raw flesh of the neck. At first we had no idea who might be collaring the animals. Pranksters? Commercial fishermen? Japanese or Soviet zoologists? The answer came in 1948 from Air Force Intelligence. The collars were remnants of waterproof bags used by Japanese for the aerial delivery of food to their forces occupying Attu and Kiska Islands. Evidently a young seal would occasionally push its head through a collar floating on the sea and afterward be unable to shake the thing off past its stiff little ears.

Marine garbage became a real hazard to seals in the 1960s, when Japanese commercial fishermen began to trawl in the North Pacific Ocean and Bering Sea with large-mesh, monofilament nets. These are woven of slick, very strong fibers resembling fishline leader. When I stopped at the Pribilofs for a few days in 1974 I saw many tangled masses of webbing on the beaches, and I saw several seals with the stuff wound around their bodies.

How Old Is a Seal?

One of our plans for estimating the number of seals called for modeling, or simulating the changes in, a population which was of unknown size but for which key zoological facts *were* known. To model called for knowing the true ages of seals in certain classes, such as the pup-bearing adult female classes. But in the 1940s and 1950s we had no means of establishing a seal's age except to measure its length, and the age-length yardstick we were using had been calculated long years before on the basis of vanished populations. Now we had reason to believe that the average seal was too small for its "yardstick age." That is, a small seal was older than supposed.

This we learned by comparing the measurements of known-age (branded) seals from the period 1913–20 with similar measurements from 1941–52, that is, taken thirty years later. During the thirty years, while the seal herd had grown more than fourfold in numbers, the average body size of its individual members had shrunk. (The shrinkage should have been predictable, although we failed to predict it.) It was a consequence of crowding, a word by which I mean to include all the stress factors—psychosocial as well as tangible—that begin to press upon an individual as its herd's population grows.

So we began to look for other indices of age. We weighed the penis bones, removed from carcasses on the killing fields, of a thousand males supposed to be three- and four-year-olds. Represented graphically, the weights fell along a biomodal, or two-humped, distribution curve—information which was useful only because it showed us that more four-year-olds were being killed than we had supposed. We weighed a thousand fresh sealskins taken at random with similar results.

Then the gods smiled upon us. In 1948, for no reason except to get the best possible photograph of a seal's dentition, I had extracted the thirty-six teeth of a freshly killed animal and had arranged them neatly on a glass plate below a camera's lens. At that moment I noticed four faint ridges circling the root of each tooth. Knowing that the skull was from an individual about four years old, the thought struck me—could one ridge mean one year of life? Indeed it could, as I learned by looking at known-age teeth taken from seals that had been marked as pups. Each ridge, in fact, is the accumulation of ivory deposited in the spring of the year, when the seal is feeding most actively.

Down to the present day, zoologists, using counts of tooth ridges, have identified the ages of more than a hundred thousand seals and have applied the data in monitoring population trends.

The tooth-ridge discovery was more than a lucky accident,

BACK TO THE SEAL ISLANDS

it was a spinoff result of scientific curiosity. I have often used it as an argument in favor of paying zoologists to spend a fraction of their time simply fooling around.

We were later to find that fur-seal teeth contain even finer ridges and markings—microscopic layers of ivory representing the ten to twelve meals of rich milk which the suckling pup receives up to weaning time. The nursing episodes are unusual. The mother gives suck for about two days, then goes out to sea to feed and to refill her breasts. While she is gone, the pup, with its belly containing up to two quarts of milk, sleeps on the beach until hunger awakens it. This feast-and-famine regimen leaves permanent irregularities in the teeth.

Visitors

One of the charms of summering on the Pribilofs was that it brought opportunity to meet interesting men and women who came to see the famous Alaskan fur seals or the equally famous sea-bird colonies. G. C. L. Bertram, zoologist from St. John's College, Cambridge University, was such a visitor in 1949. As a younger man he had gone on the British Graham Land (Antarctic Peninsula) Expedition of 1934–37, where he studied Weddell, crabeater, and leopard seals. He and his wife, Kate, subsequently became world authorities on sirenians, or dugongs and manatees. In the middle 1950s he helped me to obtain a grant to study at Cambridge University (more of this later).

On the first day that Bertram went with us to observe a sealing drive he saw the natives round up the animals and drive them to the killing fields. He was disappointed, however, that the men did not use a method he had expected to see—opening and shutting umbrellas to scare the beasts into action! That picturesque method had been illustrated in Henry W. Elliott's *The Seal-islands of Alaska,* which Bertram had studied in preparation for his visit. The book, unfortunately, was an edition of 1881.

~ In 1948 a five-man team from Harvard University including a linguist, an anthropologist, a physician, and two dentists came to St. Paul Island to study the speech and the physical features of the Pribilovians of Aleut blood. They were desperately aware that only a few years remained in which Native American cultures and bloodlines might be studied. I still correspond on matters of mutual interest with one member of the team, William S. Laughlin, now at the laboratory of biological anthropology, University of Connecticut.

Bill Laughlin has become a world expert on the paleoecology of the humans who occupied the Aleutian Islands for at least twelve thousand years before the Russian fur-seekers met them and killed all but a few hundred. He and his colleagues have counted thousands of bones and shells of marine organisms from Aleut middens in order to establish what foods the people ate at various times. Living, as they did, beside the slow-changing sea, their culture remained stable for millennia.

In 1969 I asked Bill whether any full-blooded Aleuts were still alive. He replied that this question simply couldn't be answered yes or no. "But," he wrote in reply, "a few persons are lacking any assessable traits—serological, dental, or morphological—that can be attributed to non-Aleut. . . . In other words, there are still a few persons who look like their ancestors [must have looked]."

I earlier mentioned the mummy caves of Kagamil Island. Bill interprets these and other Aleutian caves as cultural museums which were very important in the daily lives of the ancient people. Each mummy preserved the spiritual power that once resided in the living body. Hunters, for example, would often seek "advice" from the departed—whom they called the dry ones. The enormous Aleut vocabulary for anatomical terms suggests that the Aleut morticians of long ago were well trained in methods of controlling the elusive human spirit. They could hold it in the body or let it out, but in all cases they had to handle and regulate it with expert care.

~ In the summer of 1946 Walt Disney sent a man-and-wife team to the Pribilofs to photograph the seals. Perhaps he recognized in the Americans of that summer a deep weariness with

BACK TO THE SEAL ISLANDS

war and a turning toward the spiritual healing that contact with nature can bring. Anyway, he assigned Alfred G. and Elma Milotte to capture on film the intimate life of a seal from birth to maturity. Al and Elma had previously owned a photo shop in Alaska and had made a number of successful travelogues and industrial training films.

During the summer, I often saw the pair at work, huddled in parkas, red-nosed in the wind, at the edge of some rookery, waiting for an uneasy mother to give birth, or for a pup to be swamped by a breaking wave, or for a cow to be savaged by the teeth of her mate, or for some other lively happening within range of the camera. The result was *Seal Island*, a drama played entirely by seals, a drama destined to create among theater and, later, television viewers a sense of our remoteness from, and yet warm-blooded nearness to, the fur seals of Alaska. *Seal Island* was a two-reeler that ran only twenty-seven minutes. It baffled theater managers, who were used to classifying films as newsreel, comedy, or feature. Nonetheless, it received in 1948 an Academy Award and its enthusiastic reception led Disney to produce six more outdoor films in the True-Life Adventure series.

Shortly after *Seal Island* was released I borrowed a print to show to the natives of St. Paul. There it played one evening and, as the story unfolded, a half-drowned pup feebly pulled its way out of the surf while a cormorant, perched on a rock above it, seemed to be nodding encouragement. At that moment I turned to look at the shining eyes of a native who sat beside me. "Gee!" he breathed. "Gee!" During his life he had watched hundreds of true-life dramas on the Pribilof shores yet had never *seen* one of them.

Following the success of *Seal Island,* Al and Elma made other wildlife films for Disney. They finally retired, bought a hilltop mini-wilderness near my present home in Bellevue, and began to produce their own nature films and books. In 1972 Al and I enjoyed working together on a story for *Smithsonian* Magazine about the extinct Steller sea cow—a beast that weighed up to seven tons and stretched to twenty-five feet. Russian hunters killed the last one in 1768 or thereabouts. Al painted the extermination scene, guided by Steller's posthumous account in "De Bestiis Marinis," and by later information on the skeleton of the animal. (Bones crop out of the eroding sands of Bering Island even to this day.)

The Black Douglas *Restored*

The fur-seal treaty of 1911 came to an end during the war, on October 23, 1941. On December 19, 1942, the United States and Canada made a provisional arrangement to protect the seals in the eastern part of the North Pacific Ocean. A new four-nation treaty was not to be signed until February 9, 1957.

The prewar need for information on fur-seal food habits and migration routes had, of course, not been met while the war was on, so in 1947 Congress appropriated $62,500 for renewed research. I was again made responsible for the field studies, to be assisted by newly hired zoologists Karl W. Kenyon and William H. Sholes, Jr.

Karl continued in sea-mammal research until he retired from the Fish and Wildlife Service in 1973. He and I have remained close friends. Bill Sholes left the sea for state fish and game research in Washington, Alaska, and California, retiring in 1977 to pursue his hobbies of backpacking into the Sierra Nevada and of growing rare native plants.

On the refitted *Black Douglas* the three of us cruised over seventeen thousand miles in 1947 on two trips out of Seattle. The first one took us to the Pribilofs and back, the second to the Pribilofs and Attu Island and back.

During the first cruise, we stopped in early June at Samalga Island in the eastern Aleutians to verify an Aleut rumor that fur seals habitually land there in summer. Sure enough, we saw about a hundred seals resting on offshore rocks. Here was the first evidence that the Alaskan seals set foot (or flipper) anywhere on land during their long winter-spring migration. Other zoologists were later to learn that nomadic seals from the Pribilofs do occasionally land on Asian rookeries, and, incredibly, in 1968 they discovered a small colony breeding on San Miguel

BACK TO THE SEAL ISLANDS

Island northwest of Los Angeles! Four of the California animals carried marks of their Pribilof birth; another had been tagged as a pup on a Soviet rookery.

We had not cruised far on the *Black Douglas* before we saw that she was too large for our purposes. Her deck was too high above the water and her turning radius too great. Although we could count seals in comfort, we could not shoot them and recover them for study. So we took the ship on a third cruise, in November–December 1948, and then turned to using chartered purse seiners only one-quarter to one-third her size.

That third and last cruise was a rough one. After eating Thanksgiving dinner in Unalaska Bay we headed southeast across the wild North Pacific on a great-circle course for San Francisco. Once, during the ten-day run, a giant wave struck the ship's afterdeck and swept away our lifeboat and a section of the rail. We saw seals almost daily, some of them a thousand miles from nearest land. As we came in sight of the Farallons and Golden Gate, Kenyon shot a seal—the only fur seal collected during the pelagic sealing days of the *Black Douglas*.

I last heard of the *Black Douglas* in 1973 from a correspondent in Pompano Beach, Florida. "A really fine looking ship," he wrote, "but in very poor condition. She is being refitted as a school ship at the local shipyard."

Going back a bit, while the ship was docked at Unalaska and we were waiting to start south, we heard that a prehistoric monster, twenty to thirty feet long, had been found on the beach by a certain Henry Swanson. I sent one of the zoologists (who shall remain nameless) to interview Henry. He staggered back to the ship around midnight in a most deplorable state, having been obliged, he said, to swap drinks with Henry at the Blue Fox Tavern until he had drawn out the full story. Henry later brought us the head of the beast, its ugly black snout protruding from a barrel. We were glad to have it, for it was an uncommon specimen of Baird's beaked whale.

During the summer of 1947 we spent several weeks on the Pribilofs, mainly to resume the marking of pups. The percentage of marked ones returning later would theoretically give us a fix on the number that had been born in 1947. With the help of a native crew we rounded up and tagged 19,183.

We also rigged up a balloon camera and sent it aloft on a 1,200-foot line above a rookery, hoping to get photographs on which the images of seals could be counted. The camera swung and twisted fifty feet below two helium-filled Weather Bureau balloons. Defeated by misty weather and gusty winds, we eventually abandoned the experiment. Kenyon and Sholes would have quit sooner, had they not recognized my stubborn pride of invention in the thing.

We carried a few live seals to Seattle aboard the *Black Douglas* and forwarded them by train to the San Diego Zoo, where trainer Benny Kirkbride taught one of them—King—to perform. King became a reasonably good actor, although he could never seem to balance a ball on his pointed nose as well as could the California sea lions on their stubbier ones. King died five years later.

18. Author and assistant Robert Z. Brown (*right*) test an aerial camera for use in photographing the Pribilof fur-seal rookeries, St. Paul Island, Alaska, 1947. (Photograph: Karl W. Kenyon)

BACK TO THE SEAL ISLANDS

A Breakthrough in Counting Seals

From the start of our studies in 1940 we had often been asked, "Why don't you try to count the seals from the air?" Aerial reconnaissance had already proved its value as a technique in the censusing of ducks, deer, caribou, and other game species. Even seals. As early as 1927, Sergei Dorofeev had counted harp seals on the White Sea ice, while Canadian hunters in the same year had used a small skiplane for spotting harp seals off Newfoundland.

So in 1945, near the end of the war, we asked the Navy at Adak for help. They sent an amphibious plane (PBY) and a photographer to take experimental shots of the Pribilof seals. I sat beside the pilot while the photographer aimed his camera through a hole in the floor. Unfortunately for the success of our mission, three admirals bent on sightseeing came along for the ride. Conditions were not the best for experimental photography and the resulting pictures could not be used for census purposes.

But we made a breakthrough in 1948. On two bright, cloudless mornings in July, Kenyon and I photographed all the rookeries from the air. He chartered a twin-engine land plane equipped with a standard aerial camera and, accompanied by a commercial pilot, flew from Anchorage to St. Paul Island. After we had photographed the St. Paul rookeries we flew to St. George, forty miles away. I felt a bit nervous as we crossed the open sea in a land plane, watching the gulls and murres "flying backward" past our windows. There would have been no chance of rescue had we gone down.

While the photographing was underway, Bill Sholes was on the ground counting seal pups on sample rookeries. The counts provided ground-truth information which we later used, along with area measurements taken from photo enlargements, to estimate that 580,000 pups had been born that summer. The 1948 survey was a smashing success, due mainly to Kenyon's expertise and to our luck in having the Bering Sea clouds part for two successive days at the peak of the seal-breeding season.

By 1949 we were growing desperate for statistical help with our census studies so we called on Z. William Birnbaum and his

19. King, the first fur seal to have been trained, *ca*. 1950. (Photograph: San Diego Zoo)

assistant, Douglas G. Chapman, at the University of Washington's laboratory of statistical research. They cooperated fully. To Chapman, especially, the seal data offered a rich opportunity for studying theories of population change among wild animals. Later he became chairman of the scientific committee of the International Whaling Commission (1964–73) and chairman of the Marine Mammal Commission of the United States (1976–79).

Outstaying a Harem Master

The dominant, or alpha, bull seals (harem masters) guard their stations and their females with fierce and often bloody vigilance. One can easily recognize on aerial photographs the well-worn circular paths made by the bulls as each shuffles around the borders of his territory. Until 1948 we did not know how long the average bull maintains his vigil before he returns, thin in body and (one supposes) weary in spirit, to the ocean to recuperate.

So we asked Lavrenty Stepetin, a teenaged native, to stand daily watch over a group of harems on Reef rookery. He was told to record the movements of all the bulls he could identify individually by scars or color marks. We presumed that he would soon tire of his job, but he did not. Although he had been born within sight and sound of the Reef, he thoroughly enjoyed watching the endless pageant of seal life: the fighting, playing, nursing, copulating, and giving birth.

At the end of summer he reported that one bull had stayed fifty-four days, and another fifty-nine days, at his station without food or drink. (A few years later, Richard S. Peterson, zoologist from Johns Hopkins University, repeated the study and found that one old fellow chalked up a duty record of seventy-seven days! And Roger Gentry, of the National Marine Fisheries Service, learned that a typical bull loses 30 percent of his weight during the breeding season.)

Zoologists are not, however, greatly impressed by the bull seal's ability to live without drinking, for many other mammals —especially desert forms—have equal ability. Some in fact seem *never* to drink; they manufacture water of metabolism within

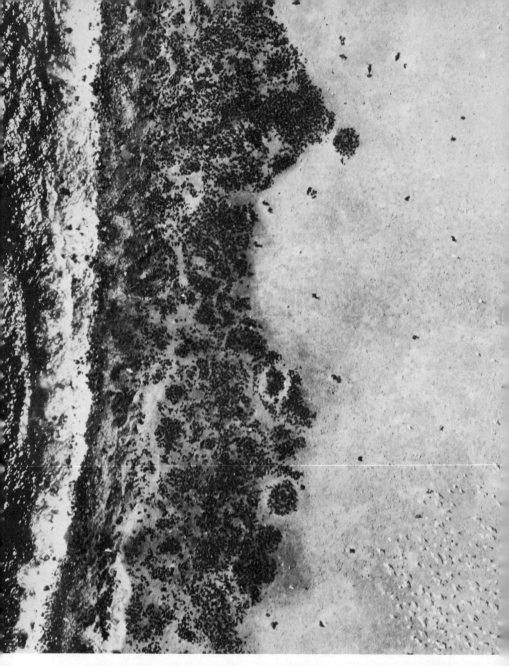

20. Aerial photograph of a fur-seal rookery on St. Paul Island, Alaska, at the peak of the breeding season, 1948. On the left is the white surf of the Bering Sea. In the middle is a sandy beach covered with thousands of adult seals, mostly nursing mothers with black pups. Each harem shows as a roundish cluster of seals surrounded by a well-beaten path which the harem bull has made in patrolling his territory.

BACK TO THE SEAL ISLANDS

their bodies by the oxidation, or chemical burning, of food products. Nor are zoologists greatly impressed by the seal's ability to fast. Very obese humans have been starved under medical care for up to 249 days.

Lavrenty's success as an observer prompted us to plan increasingly elaborate studies of the daily procession of events on a typical seal rookery. We were aware that, for a century and a half, men had stared at the seals *en masse,* grasping the broad outlines of their behavior but not comprehending its clockwork timing. So, when Kenyon and I landed on the islands in 1950 we began to photograph, at five-day intervals from May through September, a small corner of Kitovi rookery. We dragged an abandoned Army toilet to the edge of a cliff above the rookery, anchored it securely against the sea winds, and used it as a camera station. What we were after was a series of photos that would depict the arrival of seals in spring, their summer breeding, and their gradual exodus from the islands in autumn. The series would resemble one of those time-lapse movies of a rosebud unfolding or of a pea-vine tendril sweeping the air in search of support.

Our photos subsequently attracted the attention of an animal behaviorist, George A. Bartholomew, at the University of California. He spent the summer of 1952 observing the seals of Kitovi, while Dick Peterson followed him in the summers of 1961–63. Between the two, they answered the pressing questions of How many? When? and Why? with respect to the breeding animals. They found, for example, that the ratio of copulations to females is about 1 to 1.3, or that some of the females mate more than once during the few days each year while they are in heat.

I remember Dick's study of twinning in seals. The incidence of twin fetuses is known to be very low—about fourteen in ten thousand. While he was watching seals from the Kitovi observation tower Dick saw a female give birth to twins, the first-ever actual observation of a multiple birth among pinnipeds. The new mother responded to the first-born of the pair, possibly because it had the louder voice, while rejecting the second-born, which never nursed and soon wandered away. Later in summer, Dick saw another twin birth and again the mother rejected one

21. An observer clocks the movements of fur seals in order to prepare a calendar of events on the breeding ground, or rookery, 1962. (Photograph: Richard S. Peterson)

of her pups. To his surprise, he witnessed a third twin birth, this one to a female being held in a temporary cage. Here the mother nursed both pups, but when the little family was released on the rookery it stayed together for only a few days. The question of whether a fur-seal mother can successfully raise two pups to weaning age remains unanswered, although I've no doubt that some young zoologist in the Dick Peterson tradition will eventually answer it.

New Knowledge of Sea Otters

The fur-seal treaty of 1911 had given protection both to fur seals and to sea otters, with the distinction that sea otters could not be killed or taken captive for *any* purpose. As a consequence, zoologists were handicapped in studying sea-otter anatomy and physiology until after World War II, when the old treaty was replaced by a more permissive agreement.

From 1947 to 1949 we zoologists in the *Black Douglas* group assembled a large collection of skulls and skins from sea otters dying of natural causes. It was our good fortune to meet Elmer C. Hanson, an Army civilian who was stationed for several years on Amchitka Island. He volunteered to gather sea-otter remains from the beaches there and forward them to us in Seattle. Before he was transferred, he had salvaged parts of more than a hundred otters and had taken measurements of many entire carcasses. Our collection became the largest of its kind in the world. Hanson shipped the specimens dried, or frozen, or wrapped in surplus weather balloons. I report the balloon technique without recommending it for, en route to Seattle by ship, the balloons would slowly inflate from gases of decay. We had to open them carefully outdoors and downwind.

Hanson was a prime example of the born naturalist who, while making a living in some unrelated field, develops into a self-taught amateur scientist. (I shared an office at Cambridge University with another such person, an expert in dog physiology who gained a living at ship building.) The history of zoology is rich in accounts of the contributions made by the Elmer—and the Elmira—Hansons of the world.

ADVENTURES OF A ZOOLOGIST

Although we did not imagine that our sea-otter bones would ever figure in a legal decision, they were to do so in 1977. The Fish and Wildlife Service decided that the sea otters of the California coast were a "threatened species" under the Endangered Species Act. That decision was based partly on slight differences in skull measurements between Californian and Alaskan animals. The differences were cited as justification for classifying the Californian animals legally, as well as zoologically, unique. The decision touched peripherally on whether the Californian animals deserve to have their own Latin name *(nereis)*. Some zoologists said yes, others no.

I found the whole argument depressing. There exists *no* wildlife population that isn't unique and deserving of human concern, whether it has or has not been christened in a Latin ceremony. Systematic zoology should be practiced without regard for its legal or political consequences. When David Starr Jordan, in 1898, pronounced the Alaskan fur seals a distinct species *("alascanus")*, he did so on patriotic grounds, for no zoologist since has been able to find the slightest difference between the American and the Asian seals.

From 1951 to 1959 the Fish and Wildlife Service (and its splinter agency, the Bureau of Sport Fisheries and Wildlife) made five unsuccessful attempts to transplant sea otters to islands where the original populations had been hunted to extinction. In one heartbreaking attempt, captive animals were tranquilized before being air-freighted from the Aleutian Islands. All died, supposedly from the combined effects of the drug and exposure to thin air in the nonpressurized plane. Other groups died either of overheating en route to, or of chilling upon arrival at, their foster homes.

Karl Kenyon did, however, succeed in 1955 in bringing one young female otter, Susie, to the Seattle Zoo, where she lived for six years. There she spent endless hours grooming herself—rolling in the water of her pool, patting and squeezing her fur, blowing air into it, and licking it. Intrigued, Karl asked me to help him record her activities at one-minute intervals for a day. During the fifteen daylight hours of an August day, Susie spent 48 percent of her time grooming, 24 percent exercising or playing, 19 percent resting, and 9 percent feeding. (Male chauvinists

22. Sea otters spend nearly half the daylight hours grooming their fur. Photographed in 1969.

will claim to be unsurprised that a female should spend nearly half her time preening. However, to a sea otter, the problem of keeping clean—which is really the problem of maintaining the insulating quality of the fur—is unrelated to sex.)

On January 1, 1960, the new state of Alaska assumed responsibility for her own wildlife resources, and from 1965 to

1972 her zoologists transplanted about seven hundred sea otters. They captured the otters in fish nets, penned and fed them for a while in clean, cold, circulating sea water, then shipped them quickly and directly in suitable planes.

~ By 1948 I had been in civil service ten years and the profession of wildlife management in North America was ending *its* tenth year. I count from 1937, when the Wildlife Society, an organization that now includes seven thousand members, was founded. One historic milestone in the growth of the profession was the publication of its first comprehensive textbook, *Wildlife Management,* by Reuben E. Trippensee, 1948 and 1953. Another was the publication in 1948 of "Outlines for the Study of Mammalian Ecology and Life Histories," by Walter P. Taylor. Foreseeing the importance of an ecological approach to zoology, Walter wrote, "there is but one great system of matter and energy, and studies of restricted parts of the system are but for convenience. The truest, most adequate, and most satisfactory approach is through the entire biotic community." During the years when Walter was a zoologist in the Biological Survey (and later the Fish and Wildlife Service) he was regarded by certain colleagues as academic and impractical. Perhaps . . . yet I think that most of his critics came around eventually to sharing his vision of the wholeness of life.

7

A New Fur-seal Treaty

IN the winter of 1948–49 and again in the spring of 1950, the Fish and Wildlife Service sent Ford Wilke to Japan to observe Japanese methods of hunting fur seals. Harpoon boats *(tsukimbo-sen)* were, and still are, used in hunting seals, porpoises, sharks, and sunfish—animals that can be shot or speared from the pulpit of a small vessel. Ford gathered data on 475 fur seals in the harpoon-boat catches and pioneered the methods of pelagic fur-seal research which the Service was later to employ on a much larger scale. To everyone's surprise he learned that large numbers of young Pribilof-born seals spend the winter off Japan, and that they feed on squids and small schooling fishes (especially lantern fishes) rather than on species of high commercial value.

Our State Department, now having at hand both Ford's data and the postwar data from the Pribilof and *Black Douglas* investigations, decided that the time was ripe for a cooperative international study of the oceanic life of the seals as a prerequisite to negotiating a new treaty. On April 16, 1951, William C. Herrington, a fisheries zoologist working for State, proposed such a study. Beyond being an extension of our own earlier studies of seal feeding habits and migration routes, the new study would be international. Herrington foresaw the diplomatic, as well as the scientific, rewards to be gained from it.

So, early in 1952 the United States, Japan, and Canada launched a pelagic investigation. The Soviet Union, for reasons I was never to learn, refused to participate. I was placed in charge of the Eastern Pacific Expedition, sharing scientific responsibility with James I. Manzer, a zoologist from the Fisheries

Research Board of Canada. We chartered four purse seiners and their crews—two for a spring cruise to southern California and return to Seattle, and two for a summer cruise to the Gulf of Alaska and return. By the end of the summer of 1952, the zoologists of the three countries had collected and examined more than three thousand seals.

Perhaps my narrative at this point is pressing too hard on the *killing* of seals. I know that killing for research is offensive to some but I defend it when it is done humanely and when there is no other feasible way to obtain essential anatomical data. Autopsies of the seals taken in 1952, for example, contributed to scientific knowledge in the following ways.

Their genital tracts showed us that Asian seals had a pregnancy rate of 80 percent; American seals only 69 percent. That difference was evidence of the difference in crowding of the two populations; it was evidence that the Asian population was still growing while the American population was not. (We could not have obtained this information by studying live seals on land in summer, for only the highest rates of pregnancy can be expected in maternity wards.) The genital tracts showed also that the fur-seal fetus undoubtedly implants, or attaches to the uterine wall, in November, four months after copulation. This was new information.

The stomach contents of the seals showed that the animals are nonselective feeders, taking many kinds of fishes and squids. Their diet is now known to include more than a hundred species.

Their skull measurements established with certainty that Asian seals and American seals are of common stock; they can't be separated as Eastern-born or Western-born.

Counts of those which were wearing flipper tags enabled us to estimate that 30 percent of the seals feeding in waters off Japan in spring were Pribilof-born.

The foregoing and other facts established zoological grounds for negotiating a new treaty. They helped the governments of the North Pacific nations to answer, especially, the questions: Who "owns" the seals? What national waters do they frequent? What is the probable impact of their feeding habits on

A NEW FUR-SEAL TREATY

national commercial fisheries? The joint report of the 1952 investigation was published in 1955 after it had been searched minutely for political implications.

I don't wish to leave the impression that research on sea mammals *must* be lethal. It is, in fact, moving steadily in the direction of research on live, undisturbed specimens. Zoologists are, for example, mapping the migration routes of killer whales in Puget Sound, and of right whales and dolphins in Argentinian waters, by photographing and subsequently following individuals having distinctive natural marks. Students at Evergreen State College, on Puget Sound, are finding out what seals eat by examining fish otoliths in seal scats recovered from the beach. The tiny otoliths—or ear bones—resist digestion and are characteristic for each species. To date, the students have identified more than twenty kinds of fish in the diet of the seals.

Outer-space scientists with their dreams of cities in the sky are scarcely more ingenious than some zoologists I know. One is tracking whales acoustically by their individual voiceprints, or "signatures." Another is collecting the stomach contents of porpoises by catching the animals alive, administering a gentle emetic, and returning them to the sea. (Here I think of Mark Twain's fly crawling from the inkwell "alive but discouraged.") Another zoologist is shooting slender core-samplers, each attached to a recovery line, into the skin of whales. He withdraws a plug of tissue, examines it in the laboratory, and identifies, from its chromosomes, the sex of the whale.

I recently heard Marianne Riedman, zoologist at the University of California at Santa Cruz, explain how she milks an elephant-seal. She scouts along the beach until she locates a dozing female with swollen black nipples. She sneaks up to the mother and quickly presses a milking stick—a sort of suction tube—against a nipple, thereby getting up to an ounce of milk before the donor wakes in alarm. Marianne says that elephant-seal milk is the richest of any known—averaging 55 percent pure fat.

ADVENTURES OF A ZOOLOGIST

The Treaty Is Signed

The long-awaited North Pacific Fur Seal Conference opened in Washington on November 20, 1955. The Soviet Union, unwilling in 1952 to participate in joint research, now joined Canada, Japan, and the United States at the treaty table. I attended it for a week as an observer-expert, then returned to my office, for it promised to drag on for many months. Not until February 9, 1957, did it give birth to the Interim Convention on Conservation of North Pacific Fur Seals.

To me, naive in diplomacy, the conference was a fascinating mix of politics and zoology. At one point the discussion turned to the best method of harvesting seals. The Japanese delegate was arguing in favor of pelagic (as distinct from land-based) sealing. Killing an adult female at sea, he declared, usually means the loss of only one other life—that of her fetus—while killing a female on land means the loss of two other lives—those of her nursing pup and of her fetus (the latter still unimplanted and microscopic but arguably a life). The American delegate countered that, if net energy were the only consideration, the best time to kill her would be immediately after she had replaced herself, that is, after she had raised her pup to independence. The discussion died in babble about the futility of deciding the "best" time to bring a mother seal's life to an end.

On another occasion a Japanese specialist in population theory was arguing for a reduction of the Pribilof herd. He covered a chartboard with figures and curves trying to persuade us that, at a future time when a certain curve would drop below its X-axis, the rookeries would display more bulls than cows! Well, though he could change polygyny to polyandry on a chartboard, he could never do so in the real world.

The convention of 1957 continued the ban on pelagic sealing, a wasteful and nonselective method of harvesting seals. It set up a six-year cooperative research program in which zoologists of the four nations were to work side by side in the field and were to share data resulting from their independent studies. It specified that the "have nots"—Canada and Japan, which do not own seal islands—were to share the profits of annual seal harvest with the "haves"—the Soviet Union and the United

A NEW FUR-SEAL TREATY

States. It established a four-person North Pacific Fur Seal Commission. That body held its first meeting in Washington in November 1957.

The convention (as amended since 1957) is regarded both by diplomats and by zoologists as a model of compromise among nations, any of which could have, in the absence of an agreement, exploited the fur seals to the point of commercial extinction. There are some today who would prefer a treaty concerned more with the values of living seals than of dead ones —but that is another matter.

Starting in 1958, I served as a scientific adviser to the North Pacific Fur Seal Commission. During one advisers' meeting in Seattle on February 16, 1962, the Soviet delegate, Sergei Dorofeev, dropped dead beside us. We felt his loss keenly, not only because we liked him as a person but because he knew more than anyone else in the world about the seals of Soviet waters, having studied them there since the 1920s. The Soviet Embassy sent a man out from Washington; there was an autopsy; it confirmed that our friend had collapsed from an overburdened heart.

Aiming for Maximum Yield

Soon after the end of the three-nation expeditions of 1952, I had abandoned fur-seal research for a five-year period (1953–57) to carry out administrative duties for the Fish and Wildlife Service in Colorado. I returned to Seattle in 1957 to find that research was now aimed almost exclusively at discovering the level of maximum sustainable yield (MSY)—the regulated level of seal population at which the greatest number of seals could be killed for their skins. After I left Seattle, Ford Wilke carried on as director of sea-mammal studies for the Fish and Wildlife Service. He soon gathered around him a staff of men, all of whom were zoologists familiar with fish-population dynamics. To predict the MSY of the seal herd they first looked into its history. They located a baseline year in the 1930s, when the herd had been climbing rapidly from near extinction, then drafted a scheme to reshape the herd after that model year.

First they had to reduce the size of the herd, and the quickest way was to kill females—those animals that, for more than a century, had been "sacred cows" in fur-seal management. So, in the years 1956 to 1965, the Aleuts killed about 230,000 cows of breeding age. No one has yet come forth with an estimate of the cost of that killing in terms of the pups made orphan and consequently dying of malnutrition.

~ I wonder, Did the end justify the means? On rational grounds, the answer is perhaps yes; on ethical grounds, no. The decision to kill cows was made by the same officials in Washington who, in the 1950s, directed the extermination of the native St. Paul foxes—a zoological subspecies—because their pelts were of low market value.

After the seal herd had been cut back it did *not*, to the planners' dismay, produce more sealskins. Instead of a predicted 80,000 to 100,000 skins annually, it produced only 47,000. What had gone wrong?

One part of the answer seems to be that the planners used a model which, although applicable to fish populations, was not applicable to mammals. Also, an unexpected growth of Japanese and Soviet commercial fishing efforts in northern waters was beginning to leave fewer fish for the 300,000 nursing mother seals feeding in summer near the Pribilofs. A third possible answer is that rising concentrations of lead, mercury, and chlorinated hydrocarbons in the sea were insidiously depressing the growth of the herd. These poisons have been identified in seal tissues, although their effects (if any) on the seals and on the food organisms of the seals have not been measured.

Visit to a Russian Seal Island

The fur-seal treaty of 1911, having made no provision for continuing research, was blind to the certainty that the North Pacific seal populations and their ecosystems would change with the passage of time. The new convention of 1957, however, provided both for national research and for international exchange of zoologist/observers. As one conse-

A NEW FUR-SEAL TREATY

quence, I was sent in the summer of 1960 as an observer to Ostrov Tyuleniy (or Seal Island, or Robben Island), in the Sea of Okhotsk.

Claimed by Japan as a prize of the Russo-Japanese War of 1904–1905, the island was reclaimed by the Soviet Union after World War II. It is treeless, only four-tenths of a mile long, reeking with the odors of fur seals and nesting sea birds, and swarming with kelp flies. My interpreter, Eugene M. Maltzeff, and I were the first Americans to land there since Stejneger's visit of 1896. In Stejneger's day, lawless sealing had reduced the number of harems to seventy, whereas in 1960 we saw about a thousand.

We arrived off the island toward dusk on a big refrigerator ship, the *Zelenogradsk,* having dined at the captain's table on king crab and cognac. Two island workmen met us in a rowboat and landed us on the beach, where Dorofeev was waiting to greet us. He led us up the hill to our cabin, where a second dinner of king crab and cognac was spread! We saw no tactful way out; we ate the second dinner.

The Soviet zoological staff included Dorofeev and two younger persons, V. A. Bychkov and Galina K. Panina (a woman). They explained the details of their research and staged for us a demonstration of pup tagging. We judged that the quality of Soviet research was about the same as ours on the Pribilofs. We did wonder why the island workmen gathered up and dumped at sea the carcasses of four thousand seals found dead of natural causes. Perhaps they were expressing the great importance placed by Russians on neatness. Witness the absence of litter on their metropolitan streets.

The toilet on Seal Island was a curious structure—a floating one-holer that rose and fell with the tides. I suggested that it be stocked with survival rations in case a sudden wind should snap its moorings.

Ostrov Tyuleniy is a federal *zvyerosovkhoz,* or animal collective farm. Its population is swelled in summer by the arrival of twenty or more fishermen who harvest the sealskins. In winter, when the little island is surrounded by broken ice, only a few soldiers remain. All supplies, including fresh water in barrels, are brought from Sakhalin Island.

23. Zoologist V. A. Bychkov collects fur-seal teeth, indicators of growth and age, on Ostrov Tyuleniy Island, U.S.S.R., 1960.

At the time of our visit in late summer the common murre, or guillemot, chicks were leaving their nests, tumbling down the shaly cliffs to enter the sea for their first bath. We heard their gabble day and night, some of it rising through the floorboards of our cabin. We saw on the seal rookeries the flattened bodies of many chicks that seals had blindly rolled or trod on. The only other nesting sea birds on the island were a few hundred Pacific kittiwakes.

Although the 1960 sealing season was past, Dorofeev explained the method of sealing. Workmen crawl through two tunnels in the rookery soil and emerge near the water's edge, thus being in a position to drive the surprised animals inland.

A NEW FUR-SEAL TREATY

Once they are inside a wooden corral, the seals are either shot or clubbed. We saw the remains of a peculiar invention whose function Dorofeev explained, smiling. It was a huge sieve made of poles and designed to sort the seals by size as, prodded by workmen, they scrambled along its length. It must have been invented by an efficiency expert in Moscow who had never seen a mob of frightened seals in action. Anyway, it was a flop.

One morning we watched Dorofeev and his crew weigh four hundred seal pups as a means of estimating their body condition. The weight factor is used in estimating the chance of survival, or life expectancy, of each year's pup class. Men carried pups by their hind flippers to a platform scale. One recorded the sex of the pup; a second dropped it into a keg open at both ends and resting on the scale. A third noted its weight and emptied the keg by a sliding motion which dumped the pup on the ground. Crude but effective.

After a stay of two weeks, Maltzeff and I left Seal Island on September 13 on a refrigerator ship, transferred near Vladivostok to a Soviet lumber freighter, landed at the tiny Japanese village of Maizuru, then traveled by train to Tokyo and by air to Seattle. Here we were swabbed down by an FBI man whose eyes showed life only when we described the sex life of the seals.

Final Years in the Seattle Laboratory

During my final years in the sea-mammal laboratory in Seattle I enjoyed freedom to study aspects of fur-seal biology which were not directly related to censusing or to increasing the yield of salable skins. Ford Wilke, lab director, respecting my interest in microscopy and photography, let me explore, for example, the anatomy of the fur seal's hairy coat and its teeth. Helping me with technical aspects were University of Washington specialists George F. Odland (dermatologist) and Bertram S. Kraus (orthodontist). After working with human clients, they found fur seals a refreshing change.

The seal is born with a pelage of short hairs like those of a black fox terrier. After three months the pelage molts into a unique silvery coat consisting of an overhair layer (for protection) and an underfur layer (for warmth). The underfur layer contains nearly 300,000 silky fibers per square inch; sea water cannot penetrate it to the underlying skin. In examining the seal's body surface we found that the four breasts of the female are remarkably streamlined in shape, an adaptation in the direction of reducing drag during swimming. The mammary gland is a thin, broad apron stretching from the ankles to the armpits and partway up the sides of the body.

In our study of fur-seal teeth we were fortunate that the Seattle lab had, in past years, saved the skulls of hundreds of seals of known age. But to understand the prenatal growth of the teeth, Bert Kraus and I needed a few skulls especially prepared, so we asked our colleagues to collect fetuses at sea and hold them frozen. Later we put the thawed fetuses in a metal box filled with hungry carpet beetles *(Dermestes)*. The beetles, Lilliputian butchers, picked the flesh and cartilage from the jaws, exposing teeth as small as a pinhead.

We learned that the seal at birth is precocious with respect to its teeth as well as to the rest of its body. About half of the pup's *permanent* teeth have already cut the gums at time of birth. Precocity has survival value to an animal whose young are dropped in the open air and in all kinds of weather.

Toward the end of the tooth study, Colin Bertram sent me the five-inch tusk of a dugong he had collected in northern Australia. After preparing a smooth section of the tusk, I was able to see through a microscope hundreds of rhythmic lines in the ivory, some of them finer and others coarser, in the right numerical ratio to be daily and lunar-monthly increments. I concluded that, although dugongs live in tropical waters where they experience little change in ambient temperature, they doubtless make daily forays to and from their submarine pastures, and a record of their travels is left behind in the substance of their teeth.

Growth layers in the permanent tissues of animals have not, I believe, been fully explored. The life of every animal runs an uneven course, marked by stressful times and easy times, by periods of illness and of health. It seems only logical that evidence of an individual's past—that is, cell fragments, or crystals, or other "dead" structures—should persist in readable form.

Also barely explored are the biological clocks of animals. These seem to be regulated by a combination of external signals (e.g., length of day) and internal (e.g., amount of seasonal body fat). The timer is surely an electrochemical agent in the brain, but where is it situated and how does it work?

Virility of the Bull Seals

In the spring of 1965 I had grown restless. I had not been up to the Pribilofs for several years. I seemed to hear the quiet voice of Olaus Murie, "When you're stale, get out and live with your problem." So I went to the Pribilofs in June to help Ancel M. Johnson, of our Seattle lab, study the behavior of the bull seals during their breeding season. My special task would be to measure, by means of testes examinations, the virility of the bulls.

Each year in May and June about ten thousand adult males, seven years old or older, stake out their individual territories along the beaches. Several weeks later the pregnant cows begin to arrive, arranging themselves in harems around the bulls. I use the word *harem* because it is the traditional name for the unit of cows-with-bull that one can see by the thousands on the breeding grounds. Actually, the bull's attachment is stronger to the circular plot of ground he guards than to the "wives" who share it with him. He may or may not—depending on his mood and her sexual condition—try to recapture a cow on the point of deserting his harem.

Our objective in 1965 was to learn the age composition, the annual rate of mortality, and the annual rate of replacement (or turnover) of the bull class. These vital statistics, along with others, would be worked into computations by Ancel with the ultimate objective of answering the question: How many young male seals can safely be killed from each year-class and how many should be spared to become future harem bulls? The study was launched because there seemed to be too many idle bulls roaming the beaches outside the rookeries. Perhaps we were sparing (in its commercial sense) too many young seals.

Working with a crew of native St. Paul Islanders, we shot

249 bulls and examined 156 others that had died of natural causes, then had to carry their bodies, some weighing four hundred pounds, through the rookeries to the safety of the upper beaches. Seven pallbearer-zoologists cooperated to carry the heaviest bulls over the slimy rocks. We hoisted each body on a tripod and weighed it with a beam steelyard. Ancel extracted a canine tooth for age determination while I recovered the testes and pickled them in formalin for virility determination.

By the end of July we knew that competition among the bulls for breeding space is much fiercer than we had supposed. Soon after we had shot one bull, another, from the rookery fringe or from the water's edge, would occupy his vacated territory. Half the vacated territories were reoccupied within twenty-four hours, while occasionally an eager bull would rush in within minutes, even before we had had time to remove the original landholder.

More than half the bull specimens proved to be nine to eleven years old—evidence that a prime animal can expect to enjoy (if that is the right word) a breeding life of no more than two or three summers. The average yearly mortality among adult bulls is a whopping 40 percent, largely the result of fighting. At the end of summer, after Ancel had made his computations, he concluded that "commercial use of young males could probably be increased without adversely affecting the productivity of the herd."

Back at the Seattle lab, I photographed thin sections of the bull testes, enlarged the prints to 100 diameters, and measured the average area occupied by the sperm tubules. The tubules (where the sperms are produced and stored) are, in all mammals, largest during the peak of the breeding season. I found no significant variability in the 1965 samples; no difference between the testes of the idle and the active bulls. I concluded that the idles were constrained from breeding by psychosocial, rather than anatomical, factors.

During that summer among the bulls of St. Paul I was again impressed, as so often before, by the great difference in body size between the sexes. The average bull outweighs the average cow by 4.5 to 1. That difference is the greatest among mammals, although it is approached in another polygamous sea mammal,

A NEW FUR-SEAL TREATY

the sperm whale, where the ratio is about 2.5 to 1. Evolution of body size in the fur-seal male proceeded in the direction of a more able fighter, which in the social world of the seal means a more successful impregnator. But evolution was able to advance only so far in that direction without being checked. At each point in geologic time when the bulls of a novel bloodline became so heavy as to injure their own cows during copulation or to crush their own pups by trampling upon them, body weight became a disadvantage and the novel bloodline disappeared.

Here one has a classic example of Darwinian natural selection or, more accurately, natural rejection.

8

Teaching

DURING the seven years shortly before and after I retired from civil service in 1969, I taught courses at the University of Washington and at the College of the Cayman Islands. But before I describe those excursions into academia I ought to explain, as best I can, how I became drawn to teaching.

I had taught informally as a ranger-naturalist in Mount Rainier National Park from 1930 to 1934. I had been associated with the University of Washington as a teaching fellow in zoology from 1932 to 1936, as lecturer in forestry from 1938 to 1942, and as lecturer in oceanography from 1942 to 1954. Most important, I had early detected within myself a need (perhaps because of the trace of Germanic blood in my veins) to categorize, to make lists, and to formulate definitions. This drive to systematize facts and ideas stems from my need to understand them. I must first put facts and ideas into some kind of order before I can grasp them.

Organizing, Defining, and Compiling

As a prelude to my account of the teaching years I offer two examples of efforts to organize, define, and compile.

I attempted as early as 1952 to define the field of sea-mammal research. The result was an article entitled "Outline for Ecological Life History Studies of Marine Mammals" and was one of a series written by specialists, each of whom described the research methods applicable to a selected group of organisms.

[120]

TEACHING

The peculiarities of sea mammals, I wrote (substantially) are the evolutionary consequences of three challenges posed by the marine environment to those prehistoric land mammals which first ventured into its waters:

The challenge of *cold* has been met by adaptations of four kinds—insulation (warm fur or blubber), high basal metabolism (an active life and a lusty appetite), low surface-to-volume ratio (large body), and precocious young. These mechanisms for keeping warm are not equally developed in all species.

The challenge of a *fluid habitat* has converted their paws into flippers and their torsos into streamlined swimming machines.

The challenge of *three-dimensional space* has improved their diving ability (the sperm whale descending to at least 8,200 feet) and their breath-holding ability (the Weddell seal remaining under water for at least seventy minutes). It has also led to the development of astonishing powers of communication (the blue whale's voice carrying for hundreds of miles).

So it is understandable that zoologists are fascinated by the adaptations of these animals that seem to be pushing at the very limits of "mammaldom."

I should add that, by 1952, research on sea mammals was gaining momentum in many parts of the world under the sponsorship of organizations such as the Discovery Committee (United Kingdom), the State Institute for Whale Research (Norway), the Whales Research Institute (Japan), the Fish and Wildlife Service (United States), and the Fisheries Research Board of Canada. The zoologists who were engaged in sea-mammal research were mainly interested in increasing commercial yields. And with respect to fish-eating species such as fur seals, they were weighing the value of sealskins (an asset) against the market value of the fish eaten (a liability). They were also trying to save rare species such as the California sea otter, the Guadalupe fur seal, and the Florida manatee, but were too late to save the Caribbean monk seal, last seen alive in 1952. Marineland, the first of the world's great oceanariums, had opened in 1938 near St. Augustine, Florida; Marineland of the Pacific was to open in 1954. It was Marineland that supplied the two tame dolphins used by Winthrop N. Kellogg in 1958 to demonstrate for the first time the uncanny ability of sea mammals to echo-locate, or echo-range.

~ In 1963, while I was studying the seals of St. Paul Island, I noticed peculiar holes in the solid volcanic rock of Hutchinson Hill. They were smooth and regular, as though bored by a giant worm. Upon inquiry I learned that Allan Cox and David M. Hopkins, of the United States Geological Survey, had recently made the holes with a core sampler as a prelude to studying the paleomagnetism of the rock. I learned further that, many times in the past, the North Pole of Earth had been the South Pole and vice versa. A record of each reversal is imprinted in the crystalline structure of any rocks which happened to congeal at the time. It seems that Earth's magnetosphere can change its polarity quite rapidly— within less than a thousand years.

I later met Dave Hopkins and he invited me to write a chapter, "Marine Mammals and the History of Bering Strait," for a book which would deal with the ancient land bridges of Beringia. I did so, speculating on when, in the geologic past, certain species of Arctic seals and cetaceans had moved freely between the Atlantic and Pacific oceans and when, conversely, they had been barred by the bridge. (It was via that bridge that the first people—Mongoloids—reached North America at least twelve thousand years ago.) In the writing, I briefly entered the field of deductive zoology, a field that resembles those branches of geology in which age-long Earth processes must be replayed in the researcher's imagination, not in his or her laboratory.

From study of the physical differences among, and the geographical distribution of, present-day northern sea mammals I drew certain conclusions—the Ice Age series of land-and-ice masses in what are now the Bering and Chukchi seas acted as an evolutionary wedge to split the beluga, ringed seal, bearded seal, walrus, and harbor seal into subspecies. It may have split the common ancestor of the ribbon seal and the harp seal into species. Cold sea water by itself barred the transpolar movements of nine species, with the consequence that the hooded seal, gray seal, and bottlenose whale are now confined to the Atlantic, whereas the Dall porpoise, Baird's beaked whale,

TEACHING

Steller sea lion, northern fur seal, and sea otter are confined to the Pacific. The Steller sea cow was also a Pacific species until man exterminated it.

Teaching at the University of Washington

In January 1966 I found myself once again standing before a blackboard. The dean of the University of Washington's college of forestry had offered me a temporary job teaching wildlife ecology to a class of about thirty men and women. Here was an exciting prospect. I would dust off my memories of experiences with Pacific Northwest and Colorado forest animals and would discuss them in the light of modern wildlife management principles. I would talk about forests, not as tree farms, but as living, breathing systems deserving of our respect for their very vitality and wholeness. And in the act of teaching I would continue my own education.

I was nonetheless a bit nervous at the thought of facing students again. I seemed to hear a student's voice from the 1930s in a class in Zoology I: "What do you mean, Mr. Scheffer, a *higher* animal?" And what indeed did I mean by a higher animal? To that and to many other questions I could only stall, "I'll have an answer for you at our next meeting."

But in 1966 I soon found that teaching foresters was a delightful change from dissecting seals and porpoises. During the first weeks of our course we talked about the kinds of forest birds and mammals and their adaptations to ecologic niches. Then we touched on broad concepts, including reproductive strategies, mortality factors, territorial behavior, and population cycles. In the final weeks we dealt with confrontations between people and wildlife, as in those situations when bears strip the bark from plantation trees and when lumbermen destroy the nesting places of eagles and the spawning beds of salmon. By the end of the course I had decided to write a book which would describe for lay readers the profession of wildlife management. Eight years in gestation, that book was eventually published in 1974 as *A Voice for Wildlife: A Call for a New Ethic in Conservation.*

I taught the wildlife course at the University for three winters, at the end of which time the college of forestry had found a professor (Richard D. Taber) to organize a complete wildlife curriculum.

Then, in 1968, I was asked by the zoology department of the University to teach a summer course in the natural history of vertebrates. Many of the students would be high-school teachers. I taught the course in 1968, 1971, and 1972. The classes were small and relaxed; the students were friendly and cooperative.

My own instructors at the University who, around 1930, had lectured on vertebrate zoology had drawn heavily on T. J. Parker and W. A. Haswell's classic two-volume *Text-book of Zoology*, first published in 1898. Although that work is encyclopedic and is one to which I still turn for details of anatomy, physiology, and embryology, it is hardly a work calculated to excite a neophyte. So, when I began to teach, I selected the popular *Life Nature Library* of four volumes dealing with fishes, reptiles, birds, and mammals, for I felt that any student who could retain one-tenth of the information in those clearly written books would end up knowing all that he or she needed to know about the natural history of vertebrates. In our classroom discussions we emphasized patterns of animal *diversity* and *distribution,* while returning again and again to the grand process of natural selection, which has been largely instrumental in creating these patterns.

Diversity means the variety, or richness, or complexity of a community of plants and animals. It is best measured by the number of species per unit area. It is low on small, remote oceanic islands because only a few breeding stocks are able to cross the barrier of distance or of salty water. It is low in the vicinity of copper smelters and pulp mills because barriers of toxic wastes permit only a few of the hardier plants and animals to survive here. It is low in cold subalpine and subpolar regions mainly because these are poor in primary production, that is, in yield of plant-fixed carbon per unit of time. For the opposite reason, diversity is high in warm tropical forests. Thus the number of land-breeding bird species in North America varies from only twenty-six along the Arctic Slope to about six hundred in the complex jungles of Panama.

I seized upon distribution, or the spacing of living things, as a starting point for the elaboration of many zoological principles. For example, I asked, "Why do the grasslands of East Africa support wildebeest, zebra, gazelle, buffalo, topi, and eland, while the grasslands of North America support only the bison and pronghorn?" The answer is complicated but provocative. Near the end of the Pleistocene, or thirteen thousand to ten thousand years ago, at least thirty-two kinds of North American mammals disappeared. Did they do so because the climate became more extreme (hotter summers and colder winters)? The climate of Africa changed less drastically. And there were, at various times during the Pleistocene, land or ice bridges across what is now the Bering Strait. Did European mammals, crossing the bridges toward the New World, bring diseases for which the native, or locally evolved, mammals possessed no genetic immunity? Another theory proposes a Pleistocene overkill of big-game mammals by the first men who reached North America via the bridges. That theory is attacked by zoologists who point out that Pleistocene *birds*—presumably little hunted by the first men—suffered a degree of extinction comparable to that of the mammals. There remains the mystery of the horse. It vanished from the North American plains during the Pleistocene and yet, when it was introduced in the sixteenth century by man, it took readily to life in the wild on similar plains.

From this and similar questions I found it easy to move into discussion of global regions and continental drift, into biomes, habitats, and ecosystems. And from there it was a short jump to ecologic niches, or behavioral roles.

"The niche," wrote Charles Elton in *The Ecology of Animals*, "means the mode of life, and especially the mode of feeding of an animal. It is used in ecology in the sense that we speak of trades or professions or jobs in a human community."

Our class visited each summer a zoo, a saltwater aquarium, a salmon hatchery, the Washington State Museum, a small nature park, or "land for learning," maintained by the city of Bellevue, a national scenic river (the Green), and the animal rooms of the University Hospital (for insight into the treatment of experimental animals). We also took overnight trips to Mount Rainier, to a yellow-pine ecosystem at the foot of the Cascade

Mountains, and to the magnificent Arid Lands Ecology Reserve near Hanford.

This reserve, covering 570 square miles, is fascinating to biologists because its fauna and flora have remained nearly untouched since 1943, when the reserve came under control of the Atomic Energy Commission. Hunting and trespassing are forbidden. It is a well-preserved example of the sagebrush-bunchgrass, or shrub-steppe, vegetation zone which one commonly associates with grizzled cowhands and bawling cattle. By special permission our class spent a day and a night there on a hill overlooking the Columbia River. We spread our sleeping bags on the bare ground and stared at the stars until we slept. The darkness came alive with the voices of coyotes. No sound of airplanes, automobiles, or radios . . . only the whisper of sage and the music of the little wolves. In a pause in time we were part of the Old West.

We set out live traps for small mammals and caught deer mice, grasshopper mice, and pocket mice, all of which we released after examining them. We saw jack rabbits, desert chipmunks, and ground squirrels, and the gaping holes left by badgers digging for rodents. We startled a rattlesnake and let it glide slowly into hiding, shaking dryly as it disappeared. We saw overhead a Swainson's hawk engaged in hunting, and we saw on a fence post a burrowing owl lost in its own thoughts.

Deserts grow on one. They appear to be simple and quiet until one starts to dissect their working parts. Then they become awesomely complex and complete, filled with chemistry and movement and all the passions of life.

At the College of the Cayman Islands

In 1969 I learned that a Methodist minister from Portland, Oregon, was planning to found a new college on Grand Cayman Island, British West Indies. His wife, an expatriate from Cuba, shared his plans and hopes. According to its prospectus, the college would attempt to meet the higher-education needs of the Cayman Islands, including the need for liberal-arts education, for teacher preparation, for practical education, and for a

24. Students look for marine organisms on a beach of Grand Cayman Island, 1971.

center for the islands' arts and crafts. Beth and I were invited to teach without pay at the college—she to teach sociology and child development and I biology. The invitation offered a novel experience in tropical living as well as an opportunity to help a colonial people who had enjoyed few educational opportunities. Seven in ten Cayman Islanders are the descendants of blacks or Miskito Indians, among whom were slaves, shipwrecked mariners, and buccaneers.

So in September 1970 Beth and I flew to Grand Cayman. We found the islanders poor in material goods but rich in spirit; cheerful and generous. We stayed seven months, driven out in the end by lack of intellectual challenge in our work at the college, as well as by the oppressive, humid climate, the poisonous plants, and the mosquitoes. Mosquitoes! Once, from where I stood, I counted a hundred dragonflies and all were chasing mosquitoes. On a year-round basis, Grand Cayman must surely be the mosquito center of the world.

Our first night on the island was one I should like to forget. We lay naked and sweltering under a metal roof battered by rain and thunder. Tropical frogs held chorus through the dark, accompanied by *reggae* music blaring from a nearby native house. Now and then a tree toad thumped against the window screen as it struck down a dragonfly or a thumb-size cockroach. Came the dawn . . . with mocking birds, doves, grackles, and smooth-billed anis joining the jungle chorus. Our spirits lifted when we moved a few days later to a cooler house on the windward shore.

My biology class included four students—two natives and two visitors from the United States who were helping to build the little concrete-block college. I had selected a college-level textbook. When, however, one of the natives asked, "Are there really mermaids?" I saw that I was aiming in the wrong direction. Instead of studying books, we ought to be studying nature. So we began to read nature in our rich environment. There we delved for understanding of the ways in which tropical plants and animals insure their own survival and that of their bloodlines.

Grand Cayman is a low, flat limestone table draped in the greens of jungle, swamp, and farmland; outlined by shores of purest white-and-pink or delicate shades of gray. Around it a

TEACHING

coral reef flirts with the land, here circling it with slender arms, here touching it lightly, here veering out to let the seas come bursting through. Back and forth through channels in the reef the living ocean surges, changing from lilac to turquoise and back again with the changing tides and skies. On our field trips we tramped the beaches and skimmed over the lagoons in rowboats; we snorkeled with face masks in bottle-green shallows; we ventured once deeply into the damp, bug-infested inland thicket known to the natives as The Land. Everywhere we saw organisms that were new (at least to me) and everywhere we saw patterns of beauty; set-pieces arranged by nature; bits of the earth carved, cast, etched, and molded; lifeless and living things rich in texture, form, and color.

One morning we threaded our way through the jungle behind Gun Bay, climbing and dropping over a limestone path worn white by generations of men, women, and donkeys. Our goal was a manchineel tree, a species having orange fruits like small tangerines and having also a deadly reputation. The fruits are poisonous. Early mariners, shipwrecked on Grand Cayman, ate the attractive fruits and died in agony. We found a manchineel tree and cautiously examined the little balls lying on the ground beneath it. I posed the question, "Why has Nature built poison into some fruits and not into others?" If the "objective" of every plant is to spread its seeds, why should any seeds be protected by a lethal covering? My students worried the question for a while and then, with a little prompting, settled on a reasonable answer. Perhaps the manchineel seeds are destroyed by the digestive juices of mammals but not by those of birds. As a consequence, the manchineel has "learned" during its long coevolution with wild animals to produce a fruit which is both poisonous and repellent to mammals but attractive to birds.

Thus in the probing style of Aristotle's *Problems* did we question Nature, even when she seemed unlikely to answer.

On one of our field trips we entered the eerie, subaquatic region that marks the edge of a red mangrove swamp. Each mangrove tree pushes its roots into the mud and into the very sea itself, absorbing the brine without harm to its tissues. Each tree sprawls like a beast with many legs; the multiple trunk gradually becomes multiple roots descending into the mud. We approached by rowboat over a shallow lagoon and dived into a thicket of those roots. Let me quote from notes that I made on the evening of the trip:

> Flickering light reflects on our face-masks from the ceiling of air above us. We are in a hushed world, awful and ancient, among oozy stems swollen by halos of yellow and orange, fuzzed with plants and animals. Sponges, moss animalcules, gelatinous egg-masses of snails and fishes, brown and green algae—all these and countless other forms are here. A flat little oyster so thin that it seems to have no body hangs in thick clusters everywhere. (It is described in Caymanian tourist ads as "the oyster that grows on trees.") Across my vision there drifts a comb-jelly no larger than a walnut, so delicate that it slips through my fingers. As we swim slowly along, the dull brown lagoon floor unexpectedly brightens with a garden of sea anemones, some as large as dinner plates, gleaming in subtle shades of white. Here, grazing on the floor, is a sea cucumber . . . and there a sea hare the size of a grapefruit —a soft, improbable organism which reminds me of a circus tent collapsing, all waving flaps, no visible front or back. As we rise to breathe we see, above the waterline, angulate periwinkles and tulip shells, the brownish molluscs that typically feed on the green ooze of the mangrove trunks and stems.

During my stay on Grand Cayman I looked in vain for a familiar species—a dandelion, clover blossom, willow tree, English sparrow, robin, or crow. Before I arrived on the island I had supposed that I would be able to buy handbooks describing the local species. Alas! the place was called (proudly) by its Chamber of Commerce "the island that time forgot," and indeed it had been bypassed both by time and by naturalists. I soon came to recognize the more common species, if not by their Latin names then by their vernacular ones. Among the plants were belly-ache bush, fit-weed, mother-in-law's tongue, and pudding-pipe tree; among the birds, quock (black-crowned night heron), cling-cling (grackle), Auntie Katie (Jamaican oriole), and little gaulin (green heron). The Cayman lizard, or gecko, was called for some obscure reason a wood slave; the delicate, shell-less mollusk *Aplysia*, a sea cow, perhaps because it squirts a fluid when alarmed.

Surrounded by new and fascinating plants and animals I

TEACHING

photographed as many as I could. One morning, having captured a lively green frog, I placed it on a mossy rock, hid it under a small cardboard box, and focused my camera on the box. Meanwhile, two native boys had approached to watch this curious operation. After a minute, I abruptly lifted the box and, while the frog froze in the sudden glare of sunlight, I snapped its portrait. This trick is well known to nature photographers although it may have seemed voodoo magic to the children.

"What you do that frog?" asked one of them respectfully. "How you put him?"

Another of our reasons for leaving Grand Cayman was that we couldn't bear to watch the land developers carving and gouging its beautiful shores. But as faculty members of a new college trying to win sympathy and support from the Caymanians we were not in a position to protest the destruction. Perhaps we would not have been listened to anyway, for environment and ecology were concepts foreign to the local government officials.

Week after week the pumps of the dredges in North Sound sucked up the ancient sand of the sea floor and spewed it over the mangrove shore, "reclaiming" it for man's use. From the cloaca of each pump poured a flow of gray water mixed with seashells, broken starfish, sea urchins, crab legs, and waterlogged plants. Milky mud was carried by the tidal currents, eventually to smother living seaweeds and corals. Local skin divers who remembered the sparkle of the Sound now watched with dismay at the spreading blanket of ooze.

But there is hope for the Cayman shores. In the summer of 1972, after we had left the island, a group of natives and expatriates from the United States formed the Cayman Islands Conservation Association. Sparked by Nancy Sefton, a teacher, the association was soon joined by tourists, schoolchildren, businessmen, government employees, housewives, and others interested in saving the environment. By 1976 its membership had risen to about one hundred and it had published a handsome eighty-page *Guide to the Natural History of the Cayman Islands*. I don't know whether the voices of the association have carried yet to those lofty political chambers where conservation policy is set. Nonetheless, I am encouraged by the steps which the association has taken.

9

Counseling

A zoologist who specializes in the study of wild animals soon learns to respect their natural ecosystems. Even those which seem the simplest—the desert and the open sea—are mysterious and wonderful. As a sociologist cannot study humans outside their surroundings, so a zoologist cannot study wild animals outside the ecosystems within which they feed, breed, and find shelter.

Efforts in Wildland Preservation

It is a short step, then, for the zoologist to take an active part in saving wildlands—parcels of the countryside—for their scientific value. And many of us in zoology move beyond intellectual respect for wildlands to deep affection for them. Knowing that animals, given a healthy and stable environment, will maintain the genetic vitality of their species, we argue that people, too, need natural places where they can renew their spiritual vitality. So we try to preserve wildlands not only for their scientific value but for their educational, recreational (in the purest sense), and aesthetic values.

Sentiments of this kind have often been expressed by North American visionaries of wildland preservation, including George Perkins Marsh, Henry David Thoreau, John Muir, John Burroughs, Teddy Roosevelt, Gifford Pinchot, and Robert Marshall. In my own generation, Aldo Leopold was perhaps the one man most influential in shaping an American land ethic. In 1948, the year when he died fighting a neighbor's grass fire, he

COUNSELING

wrote *A Sand County Almanac,* now a classic text in the environmental movement. There he made a plea for "the preservation of some tag-ends of wilderness, as museum pieces, for the edification of those who may one day wish to see, feel, or study the origins of their cultural inheritance."

I never met Leopold. Only after I had seen the effects upon wildland of bulldozer, chainsaw, hydraulic pump, earth shovel, drill, and dynamite did I fully understand what he was saying. And I eventually came to agree with William R. Burch, sociologist at Yale University, that injustice to men and injustice to the Earth are not separate matters.

Hiking with Justice Douglas

In August 1958 I joined a group of hikers led by William O. Douglas along the Olympic seacoast of Washington from Cape Alava to Rialto Beach. It was a publicity trip. We were calling attention to the need for saving one of the last fragments of seacoast untouched by roads on the Pacific Coast of the contiguous United States. We felt that a token strip—a wilderness relic—should be preserved forever for the benefit of those willing to enjoy it on foot.

We camped at night in the shelter of giant yellow drift logs. By day we walked, and talked, and sent flocks of sanderlings into quicksilver flight above the surf. We followed bear tracks in the sand and caught a glimpse of a raccoon scurrying into the forest's edge with a crab in its mouth. We heard the undertone of breaking waves, pierced by the cries of gulls and crows. One evening, hard pressed for drinking water, we had to brush aside water-striders and floating leaves from a tea-brown, late-summer pool. We paused at the Chilean Memorial, a stone marking the common grave of twenty who had died here in a shipwreck in 1920. We paused at another stone bearing an Indian pictograph, its creator unknown.

At the end of the hike we were met by Larry Venable, a trustee of the Automobile Club of Washington, carrying a placard: BIRD WATCHER GO HOME! Douglas, smiling, shook the man's hand and said, "We had a fine trip; wish you could have been with us." The Club's next newsletter carried a front-page story:

> ELITE HIKING GROUP OPPOSES COAST HIGHWAY . . . The Auto Club was an interested onlooker last month as about 70 red-blooded American nature lovers unwittingly "desecrated" their preserve-at-all-costs wilderness with a 20th century publicity stunt that backfired . . . [I fail to get this point]. The Auto Club thinks it would be in the public interest to build a highway along the coast, making it possible for everyone to enjoy the tremendous awesome beauty. No one has the right to deny access to these areas to folks who have neither the ability nor the inclination to tramp through the brambles.

Douglas led a second trip along the coast in August 1964, from Oil City north to La Push. On this trip we missed a familiar face. A year earlier, Olaus Murie had lost his courageous struggle for life. One of his last contributions had been an editorial in *Pacific Discovery* Magazine in 1960, wherein he called attention to an ongoing Forest Service program in Wyoming aimed at producing more grass for livestock. It was carried out by dropping herbicides from airplanes at the expense of aspen, willow, sage, evergreens, beavers, antelope, and grouse.

"What we need today," Murie concluded, "is not technicians pursuing a narrow line to achieve a simple goal. . . . What we need desperately are scientists who accept social responsibility and who can envision the biotic complexity of managing land to serve many social needs."

Douglas's long campaign eventually paid off. In 1976 Congress added a 7.5-mile roadless strip to Olympic National Park. If Park policy endures, hikers along that strip will never see, hear, or smell motor vehicles nor be offended by the litter which invariably marks their passage.

William Orville Douglas—mountain climber, civil libertarian, lover of Earth—must be ranked near Leopold as one who shaped America's land ethic. He showed us that caring for the health and the stability of our globe is as truly a moral responsibility as is caring for people (or, if you wish, for their souls). When the Supreme Court ruled in 1972 that the Sierra Club

COUNSELING

lacked "standing" to challenge the development of Mineral King Valley as a ski resort in a California national forest, he demurred. Natural objects, he insisted, equally with faceless corporations should have the right to sue, and he went on to say: "So it should be as respects valleys, alpine meadows, rivers, lakes, estuaries, beaches, ridges, groves of trees, swampland, or even air that feels the destructive pressures of modern technology and modern life. The river, for example, is the living symbol of all the life it sustains or nourishes—fish, aquatic insects, water ouzels, otter, fisher, deer, elk, bear, and all other animals, including man, who are dependent on it or who enjoy it for its sight, its sound, or its life."

The notion that inanimate objects may suffer at the hand of man and consequently have the right to sue for damages had been proposed by Christopher D. Stone, professor of law at the University of Southern California, only weeks before Douglas seized upon it, saw its logic, and gave it wider audience.

I picked up the other day an old issue of *Natural History* Magazine and read an article in which the writer saw "in the automobile, good roads, and opportunity through them to get out and see the beautiful in nature, some of America's greatest antidotes to Bolshevism." The year was 1925. Well . . . perhaps. I think it more likely that, only if we *do* move nearer "Bolshevism," or at least away from land-use planning with dollar profits uppermost in mind, will we save what remains of that writer's beautiful America.

~ I have dwelt at some length on the Douglas hikes because they illustrate that all practicing zoologists are called upon from time to time to state individual positions on environmental issues having political implications. In this instance, the wilderness folk were set against the automobilists.

At other times I have been asked, for example, "Do you believe that wild horses and wild burros on the grazing ranges of the American West should be given the same protection as native wildlife species?" I have countered with the suggestion that specimen populations of horses and burros be placed on large fenced reservations where they won't compete with live-

25. Justice William O. Douglas and hikers stop for lunch at Third Beach, Olympic Peninsula, Washington, 1964.

stock. Or the question, "Goats and pigs running wild in the national parks of Hawaii are destroying native vegetation. Is the solution to encourage sport hunting?" I have answered no; partly because that solution is proving ineffective and partly because I have personal reservations with respect to killing for fun. Those who value life lightly diminish life's worth for all of us.

COUNSELING

A final thought. Although I believe that we who labor in zoology should openly state where we stand on issues of this kind, we should admit that beyond zoology our vision is unlikely to be clearer than that of nonzoologists.

The Nature Conservancy

I returned from the first Douglas hike in 1958 filled with land-saving enthusiasm and was thus ripe to welcome a suggestion, made by a neighbor, that a few of us living in the Seattle-Bellevue region organize a chapter of the Nature Conservancy. The neighbor was Charles Wesley Bovee, a then elderly man who had been Bellevue's first mayor. Fourteen of us met in Bovee's home on December 13, 1958, and, on September 24, 1960, we were chartered as the Washington State Chapter. The first officers were Ira Phillip Lloyd (a cabinet maker), Edith Hardin English (a botanist), Joseph A. Witt (a horticulturist and now director of the University Arboretum), and I (a zoologist). The chapter grew rapidly; it split in 1965 into the Western Washington and the Inland Empire (eastern Washington) chapters.

The Nature Conservancy (national) had begun in 1917 as a scientific committee of the Ecological Society of America. It became in 1946 the Ecologists Union and in 1950 the Nature Conservancy. It set aside its first preserve in 1954 and since then has been instrumental in saving well over a million acres of wild ecosystems. After acquiring title to lands, it usually donates or sells them without profit to established land-management agencies of government. In the role of broker it can act more quickly than can the average government agency.

Our small Western Washington Chapter showed its first vital sign by preserving Wildflower Acres, a 25-acre woodland near Marysville, Washington. The land had been donated by Tam and Ivah Deering, two spiritually beautiful old-timers who chose financial sacrifice in order that Puget Sound naturalists might have access to an outdoor classroom. In 1968 we transferred title to the land, along with responsibility for managing it, to Western Washington State College.

Partly through the efforts of the Western Washington

Chapter, lands and waters worth millions of dollars, although really priceless, have been saved. They include the Mima Mounds, peat bogs, cedar swamps, pristine islands in Puget Sound, river and lake frontages, heron rookeries, bald-eagle refuges, and—acquired in 1977 at a cost of nearly a million dollars—a primitive strip of Olympic seacoast, the Point of Arches.

A Secret Mission

In 1963 I was called to Washington to be screened for a secret mission. Told only that I might be sent to Antarctica, I assumed that the mission would be zoological; it proved to be diplomatic. In August of that year I found myself seated in a small, dimly lit room in the headquarters of the National Science Foundation. About a dozen men faced me around a table and questioned me. "Have you ever been drunk? . . . What are your views on war? . . . Do you have any bias for or against Russians?" These and other questions puzzled me, for I still believed that the mission was to be zoological. I tried to answer fully, admitting that, yes, I had been drunk a time or two—including that night on Seal Island when I had been obliged to eat two crab-and-cognac dinners.

Secrecy was lifted several months later. The United States Arms Control and Disarmament Agency had selected four men to inspect foreign bases on the Antarctic Continent. The three beside myself were John C. Guthrie (team leader and the State Department's expert on Soviet affairs), Michael Ivy (Russian linguist and State's policy expert), and John F. Ruina (professor of electrical engineering, Massachusetts Institute of Technology).

The mission was to be an exercise of Article Seven of the Antarctic Treaty of 1959—the landmark treaty which established Antarctica as the world's first international scientific reserve. Article Seven provides for mutual inspection of scientific bases by member nations. Lawyers in the State Department felt that our forthcoming mission would establish precedent in international law for future peaceful inspections elsewhere in the

26. Secretary of State Dean Rusk meets with the first United States Antarctic Observer team, Washington, D.C., December 3, 1963. *Left to right*: John F. Ruina, Michael Ivy, Rusk, the author, and John C. Guthrie. (Photograph: Department of State)

world—particularly behind the Iron Curtain—and would also put the world on notice that the United States firmly supports the peace-keeping provisions of the treaty.

We four on the team were briefed intensively on salient features of Antarctic history, meteorology, geology, oceanography, and biology. A talk by Ambassador Paul C. Daniels, signer of the 1959 treaty, was especially eloquent; we rose to our feet when he had finished speaking. Walter Sullivan, then science editor of the *New York Times,* gave an abstract of his 1957 book, *Quest for a Continent,* one that I still regard as among the best of its kind. We heard, in his chambers, a pep talk from Secretary of State Dean Rusk. We paid a one-day visit to CIA headquarters in Arlington, where the doorknob of our meeting room was not an ordinary knob but a combination lock. On one occasion I was escorted to the men's room by an agent who said that he was supposed to go in with me but would make an exception; he would simply watch the outer door!

Our team went down to the ice in January 1964 in a Lockheed Hercules (four-engine turbojet) from Christchurch, New Zealand. Fresh strawberries were on sale in the shops when we left, while halfway down—five hours out of Christchurch—the captain lowered the temperature in the plane and told us to change into parkas and insulated boots. During the long flight my thoughts turned often to the contrast between this easy voyage of discovery and the voyages of the early Antarctic explorers, many of whom had died of scurvy, frost, or starvation while struggling to answer the question: What is *there?*

Antarctica has been known for less than two centuries. Between 1772 and 1775 Captain James Cook circumnavigated it without seeing it. In the southern summer of 1820–21, Edward Bransfield (England), Nathaniel Palmer (United States), and Fabian G. von Bellingshausen (Russia) saw distant peaks and thus proved the existence of a continent, but it was not until 1841 that James Ross (England) penetrated the ice pack and reached the coast of Antarctica itself. The South Pole was reached on foot by Roald Amundsen in late 1911 and by Robert Falcon Scott a month later, in 1912. Richard Evelyn Byrd flew over the Pole in 1929 in a tiny airplane about like Lindbergh's *Spirit of St. Louis.* As recently as 1956, George Dufek, of the United States Navy, was the first to step from an aircraft onto the polar ice. The rest is modern history.

Shortly, we looked down on scattered icebergs and then the peaks of Victoria Land. We landed at McMurdo Airport, a permanent field on the floating ice of Ross Sea. The ice here is ten feet thick and has not changed perceptibly in fifty years.

The operational directive for the conduct of our inspection specified that we look for the following violations (among others) of the Antarctic Treaty:

> Any observations of measures of a military nature . . .
> Any evidence of nuclear explosions in Antarctica or the disposal there of radioactive waste material. [A curious item, for only the

COUNSELING

United States was disposing of such material; this from a small energy plant at McMurdo.]
Description of any limitations on U.S. Observer freedom of access to any areas of Antarctica . . .

Although I was not privileged to see the final report of our team, I was told informally that none of us had seen important violations. Certainly *I* did not.

During our twelve-day stay in Antarctica we paid whirlwind visits to New Zealand's Scott Station, to the Soviet Union's Mirny and Vostok stations, and to France's Dumont d'Urville Station (the last by overflight; it has no landing field). In three successive days we were at the three Southern poles—the geographic (Amundsen-Scott), the magnetic (Dumont d'Urville), and the geomagnetic (Vostok). Is this a record?

Vostok Station is a bleak cluster of wooden houses buried, except for their entrance tunnels, below the surface of a vast frozen plain at 11,440 feet above the sea. Here in midsummer the thermometer stood at minus 31° Fahrenheit. (The coldest temperature ever recorded on Earth—minus 127°—had been recorded at Vostok in August 1960.) The captain of our Hercules did not dare stop the plane's engines in the thin subzero air, so he flew the ship back to McMurdo and returned for us on the following day . . . or at least I think it was the following day; the sun did not set while we were in Antarctica and the Russians were keeping Moscow time, five hours different from ours. The Russians were fascinated by the ski-landing technique of the Hercules, for they themselves had never tried to reach Vostok by air. As we settled out of the blue-black sky we looked down on their figures aiming cameras at us.

It was plain that they had enjoyed few visitors. They hovered around us, plying us with tea and talk. One young fellow wearing Rasputin's beard insisted that I take a souvenir—a book from the station's tiny library. It was a greasy, well-worn copy of Bret Harte's *Selected Works,* in Russian. I took it to please him, though reluctantly, for I knew that everything at this remote spot on the bottom of the world had been brought via tractor train, each trip from the seacoast requiring two weeks.

ADVENTURES OF A ZOOLOGIST

At our McMurdo base we were treated as visiting dignitaries. One evening we dined with the admiral; on another we were guests at an officers' cookout. Here in the warmth of a quonset hut, drinking a purple concoction of grape Kool-Aid, water, and medical alcohol, we watched through the windows as noncoms fried steaks on a grill among the snowdrifts. The next evening we were flown by helicopter to the icebreaker *Atka,* and thence to Scott's last hut at Cape Evans and to an Adelie penguin rookery at Cape Royds. I became hooked on penguins at that point but was never to have further opportunity to study them. Shortly before our 1964 visit, five penguins had been flown from McMurdo to Wilkes Station and, eleven months later, three had returned after swimming at least 3,800 kilometers (2,400 miles)!

During our twelve-day mission, images flowed so rapidly into my mind that I was unable to sort them out; I moved in a state of euphoric shock. Antarctica holds seven million cubic *miles* of ice, some of it more than two miles thick . . . mountain ranges rearing above sixteen thousand feet with only their tips (nunataks) exposed . . . dry, moaning winds scouring the century-old, mummified carcasses of seals . . . the constant danger of death by freezing or fire (several men at Mirny, during the winter preceding our visit, had burned to death in their dormitory) . . . the silent huts on Ross Sea where the first explorers lived . . . the penciled message found on Scott's cold body on December 12, 1912: "I do not regret this journey which shows that Englishmen can endure hardships, help one another and meet death with as great fortitude as ever." The message has been inscribed on the Scott Monument in Christchurch, New Zealand.

Although our trip was diplomatic, I did find time to cut a square of skin from a Weddell seal's carcass, a sample that was later to prove useful when I studied the hair patterns of the world's pinniped species. And I visited the portable hut on the Ross Sea ice where Jerry Kooyman was soon to discover that the Weddell seal can dive six hundred meters (nearly two thousand feet) deep.

~ Let me move ahead to 1972. In that year a Convention for the Conservation of Antarctic Seals was signed by representa tives of the same twelve nations that had been party to the Antarctic Treaty of 1959. The convention provides for controls on commercial killing of Antarctic seals, should any of the party nations begin sealing. None has yet done so. Although the treaty of 1959 had given refuge status to all lands and ice shelves south of 60° South, it had left the high seas and the drift ice open to commercial exploitation. The news of the 1972 seal convention was discouraging to those who had hoped that the environs of Antarctica, as well as the continent itself, would remain forever unstained with the blood of animals killed for profit—forever a place of quiet beauty and wonder.

I now believe that some commercial exploitation of the living resources of the Southern Ocean is inevitable, if only because a hungry world is crying for fats and proteins. The plankton, or krill, alone might yield 100 million tons annually; a total greater than the present yield of all the world's fisheries. The best that one can hope for is conservation planning in advance of exploration—a sort of thoughtfulness for which there is little historic precedent.

Worldwide interest in Antarctica's krill, fish, oil, and minerals is quickening. The 1959 treaty provided for future meetings of the signatory nations. (These are the original twelve plus Poland, which signed in 1977.) The Ninth Consultative Meeting of the Antarctic Treaty, which ended in London on October 7, 1977, drafted the first outlines of an agreement to govern fishing and it declared a moratorium on oil exploration and extraction.

Enormous problems of resource protection remain. How will fishing rights be apportioned? What rights in Antarctica are due the nations—especially the so-called developing or Third World nations—outside the thirteen? What claims, if any, of sovereignty on the continent itself are valid? How are the wild birds and mammals of Antarctic shores to be shielded from the harmful side effects of tourism?

Antarctica is a testing ground for humankind. It is the last and largest wild, clean place left on Earth. We will shortly learn whether we can respect such places for their naturalness and love them for their power to move us. If we can, we may survive as *Homo sapiens*, the thoughtful species.

ADVENTURES OF A ZOOLOGIST

Among Walruses in the Bering Sea

Laurence Irving, a physiologist at the University of Alaska who had been studying aquatic animals for nearly forty years, phoned me in early 1968 to ask, "Would you like to join an *Alpha Helix* expedition to Bering Sea?" Would I! The *Alpha Helix* is a sleek white research vessel owned by the University of California. She was slated that year to carry a team of physiologists through the Aleutian Islands into the Bering Sea, where they would study the adaptations of animals to cold. My job would be to suggest places where specimens might be captured alive and also to provide information, during brainstorming sessions, on the zoology of the species taken. I joined the ship at Kodiak on the third of May and disembarked at Gambell, St. Lawrence Island, on the twenty-eighth.

The expedition leader, Kjell Johansen, of the University of Washington, wanted to be very sure that we would have at least one specimen for study, so he asked me to bring a seal by jet plane from Seattle. Coals to Newcastle! So I borrowed Oscar, a 140-pound harbor seal, from the Tacoma Aquarium and eventually got him into a comfortable cage on the deck of the *Alpha Helix* at Kodiak. On our first night out of Kodiak a heavy sea tore the cover off his cage and washed poor Oscar overboard. I suppose that, for a while, until he adjusted to life among wild harbor seals, he felt very lonely out there in the North Pacific darkness . . . no companions . . . no human friends from whom he could beg food.

Our cruise was the sort of adventure that zoology students dream about. I often wish that more of them could leave their stuffy classrooms and feel the bracing winds of wilderness. For a tiny fraction of the cost of sending men into outer space, and building supersonic aircraft and atomic submarines, our nation could afford to finance nature apprenticeships, or internships, for thousands of young men and women, thus enabling them to see the natural world at firsthand instead of through books, television, and museum cabinets.

Johansen's fears notwithstanding, we did capture live birds and mammals for research. To be sure, we wasted a whole day in Cold Bay chasing sea otters from a speedboat, hoping to

COUNSELING

capture a surfacing animal in an oversized "bug net." Not one could we approach before it had taken a fresh breath of air and lost itself once more beneath the surface of the bay. The next day, however, we set out a fisherman's gill net, forty fathoms long and three fathoms deep, among strands of floating kelp where we had seen otters feeding. Toward evening of the same day we revisited the net and found a fine, healthy otter struggling in its meshes, hissing as we came near. Two hours later, when we had stowed him safely in a tank of sea water aboard the *Alpha Helix,* he ate a chiton (a species of sea mollusk) and, later, crabs, herring, and filet of sole.

Physiologists David Leith, of Harvard University, and Claude Lenfant, of the University of Washington, studied the blood chemistry of a female harbor seal which we had captured while she was sleeping on a beach in Cold Bay. I watched as she lay on a table in the ship's laboratory under general anesthesia, her chest rising and falling, the ceiling lights reflecting from her wet, silvery body.

Suddenly there's an accident! The hose connection on the respirator has broken loose. Where's the ship's engineer? He comes quickly with a coil of fine wire, lashes the hose into place, and the seal resumes her breathing. Two men stand beside her head, two hold her flippers, and a fifth inserts a cannula into an artery. Twenty minutes later, the blood sampling is finished. Of the results, I learn only that the blood volume is 132 milliliters per kilogram of body weight. This is a high value; it is related to the seal's ability to swim actively under water for at least twenty minutes without dying from lack of oxygen. The comparable blood volume for a human being is about 72, and for a race horse, 110.

Dave discovered that an eighty-pound sea otter had a lung capacity five to eight times its predicted value, or twice that of a grown man! What this means is that the otter is admirably equipped to hold its breath for long periods while it is feeding along the sea floor. And in studying the anatomy and performance of a ribbon seal's lungs, Dave found evidence that, when the animal dives deeply, its lungs collapse and expel their content of air, thus preventing aeroembolism, or "the bends."

27. Zoologists transfer a live sea otter from a net to the launch of the *Alpha Helix*, Cold Bay, Alaska, 1968.

COUNSELING

Our most lively adventure on the *Alpha Helix* began at six o'clock on the evening of May 13, off Bogoslof Island. After the ship had dropped anchor in a thin, chilly fog we searched the beach through binoculars. We saw hundreds of Steller sea lions crowded side by side, arranging themselves for the breeding season. Pupping, however, had not yet begun, so we felt little concern about disrupting the rookery for several hours in order to capture an adult.

An offshore current of air carried the basso-profundo, fog-muted roaring of the animals to our ears as we prepared for landing. Ten men would slide through the surf in two skiffs and would immediately cut off a manageable "pod" of females. (But no males, which can weigh as much as a ton and are very aggressive during the breeding season.) We would take with us four snares; two rope nets, a syringe loaded with a tranquilizing drug, a respirator to be used, if necessary, to revive an overtranquilized subject, an aluminum stretcher, and a first-aid kit (human sort).

~ I open my diary to May 13 . . .

Daylight is dimming. As we approach the shore, thousands of murres and kittiwakes fall from the cliffs, complaining harshly. The sea lions stampede toward the water but we hit the beach and corner five of them in a cave. We throw the net over a smallish female while a man leaps on her back and binds her with ropes. (She will later tip the scales at 225 pounds.)

We hear shouts from the men who were left holding the painters of the skiffs. The sea has turned and the boats are rolling and taking water. I rush to salvage my camera and find it floating in a packsack in one of the skiffs. (Later I was able to clean and dry it, although poor Jan Iversen, a zoologist from Denmark, found his Leica thoroughly soaked and useless for the rest of the trip.)

The capture gang, sweating in spite of the subarctic chill, now arrive with the sea lion, trussed from snout to tail and fastened to the stretcher. Four men roll her into a skiff. Two others approach, each carrying a newborn sea-lion pup. I am surprised, for I had predicted that no pups would be found this early in summer. (They are, in fact, the first two of the season.) All of us are soaking wet. We shout good-natured curses at John Krog, from Norway, who tarries on the beach to aim his movie camera as we hold the sides of the skiffs, now pounding dangerously on the beach. By eight o'clock, without casualty to man or beast, we are aboard the ship.

We later concluded that the smaller of the two pups was a "preemie," for he lived only two days. The other was a normal female who quickly won the affection of Dave Leith and named herself *Baw*! Within thirty-six hours of adoption, Baw knew Dave's voice and rejected everyone else's. He nursed her on a hastily formulated mixture of eggs, margarine, dried milk, vitamins, and warm water. (It's curious that she tolerated the cow's milk, for natural seal milk contains no milk sugar, or lactose, and, in fact, this sugar can give a captive seal the "scours," or fermentative diarrhea.) When I last heard of Baw she was living in New York Aquarium, having in the meantime traveled with Dave by air from Alaska to his home in Boston, where she fraternized for several weeks with Dave's three sons and the family dog.

We returned to Bogoslof the next day, approaching a side of the island where there were no sea lions. We pulled from their burrows, or caught in mist-nets, a number of live puffins, petrels, and murres. (Japanese mist-nets, formerly woven from human hair, are now woven from tough black plastic thread.) We also captured a dozen petrels and a murrelet that boarded the ship at night and stayed to become the unwilling subjects of research on body temperatures.

When we reached the Pribilof Islands we captured a bull fur seal by drug syringe and carried him via truck to the St. Paul laboratory. The next day, we released him in the ocean and watched him swim uncertainly toward his old territory on the reef. When the ship reached the southern edge of the Arctic drift ice, near Nunivak Island, we took live harbor seals, ribbon seals, and walruses.

There were days when the lab of the *Alpha Helix* resembled a three-ring circus. In one corner a seal would be flapping along on a treadmill immersed in shallow ice water; in another

COUNSELING

a seal would be rising periodically to an artificial breathing hole where its breath was trapped for analysis; in a third a seal would be dozing in a corset which electronically recorded its blood flow.

One chilly morning as we cruised among the ice floes off St. Matthew Island a crewman burst into the lab. "Killer whales!" he cried. Grabbing my camera, I went on deck. Five whales were moving back and forth in the leads with sea birds fluttering above them. The whales seemed to be hunting for something—perhaps for seals which had sought refuge among the floes.

I always find it hard to describe the sensual impact of seeing a killer whale up close. The words "intelligence," "style," and "perfection" come to mind, though they fall short of the mark. On this occasion I was able to get a clear photograph of one killer, probably a subadult male, having peculiar deep scars along its back that looked as though they had been made by a wide rake. Later, in the Seattle lab, I measured the spacing of the teeth in one of our killer-whale skulls and concluded that I had captured photographic evidence of infighting among killer whales. (Shortly after seeing those killers in the wild I watched the "letting down" of Skana, a female killer whale in Vancouver Public Aquarium. Her pool is drained periodically, and while she lies quietly on the floor she is examined by veterinarians and physiologists.)

Before venturing on the *Alpha Helix* trip I had never been in walrus country. About ten o'clock on the evening of May 23 we saw a dark, huddled group of these beasts—about twenty-five young and old—resting on an ice floe. As our ship slid quietly toward them we were able to photograph them in the fading twilight.

Soon we landed at Gambell, the historic center of the Alaskan Eskimo's sea-mammal culture. Here the National Park Service has placed, on the skull of a bowhead whale, a bronze marker announcing that Gambell is a National Historic Landmark, a place where Native Americans have lived without break for thousands of years on the products of the sea—whales, porpoises, seals, birds, and fish. We descended into one of the traditional "meat holes," or cellars in the permafrost where

28. Danish physiologist Rick Tønnesen, aboard the *Alpha Helix* in Bering Sea, 1968, holds a spotted seal in a corset which measures its blood flow.

walrus meat is stored. And here we saw a modern sign half buried in the snow: AVON COSMETIC SHOP—BEDA SLWOOKO, PROPRIETOR.

History is moving rapidly at Gambell. The 1960s brought shocking cultural changes to the people's fine-tuned hunting and fishing economy. Snowmobiles belching fumes replaced the thousand sledge dogs that formerly lived in Gambell and consumed over two hundred tons of walrus meat a year. By 1970 the village could boast a single dog team, a museum piece. Still later, upon my appointment to the Marine Mammal Commis-

COUNSELING

sion in 1973, I learned that a few of the mainland Eskimos were killing walruses simply for their tusks, leaving the carcasses to rot.

~ The findings of the *Alpha Helix* expedition were not published as a set; some were released in scientific journals, others found their way into physiology handbooks, while still others were used as teaching material by the educators among us. I point to one example of a journal article. Jan Iversen and John

29. David R. Leith records blood flow, via scintillation counter, of a harbor seal, aboard the *Alpha Helix* in Bering Sea, 1968.

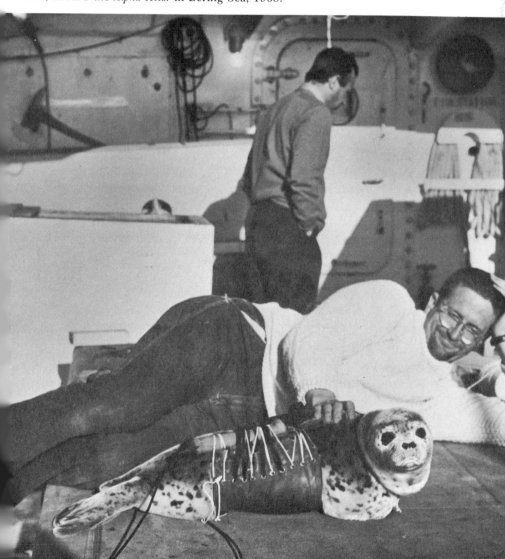

30. The blubber temperature of a harbor seal, here immersed in an ice bath, is being recorded electrically aboard the *Alpha Helix* in Bering Sea, 1968.

Krog reported on the heat production of sea otters and six kinds of seals. As measured by oxygen consumption, the resting heat production was higher than in land mammals of similar size. Especially "fast-living" were two sea otters (each weighing 88 pounds) and three young walruses (averaging 133 pounds). These burned, or required, nearly 3,000 kilocalories per day. High heat production is evidently an adaptive response to the cooling effect of the oceanic environment.

COUNSELING

Among Manatees in Guyana

On the northeastern bulge of South America lies Guyana, a small country well known to collectors of exotic tropical birds, butterflies, snakes, and fishes. Rising over nine thousand feet from its eastern flank is Mount Roraima, that rude thrust of Precambrian rock which served as the locale for A. Conan

31. Physiologists at Vancouver Public Aquarium study Skana, a killer whale, while her pool is empty, 1968.

32. Walruses rest on a floe north of Nunivak Island, Alaska, 1968.

Doyle's science-fiction book *The Lost World.* In early 1974 I learned that I was slated to attend a manatee workshop in Georgetown, Guyana. The meeting would bring together about forty zoologists and would be sponsored by the National Science Research Council (Guyana) and the National Academy of Sciences (USA). One joker at the workshop dubbed it MA-NATEE, or Man Assisting Nature to Alter Trichechus Environment Endeavor. (*Trichechus* is the generic name of the manatee.)

The manatee is an obese, grotesque mammal weighing up to half a ton. It lives in sluggish tropical rivers, occasionally venturing into the ocean, on both sides of the Atlantic. It is wholly vegetarian and may eat more than two hundred pounds of aquatic plants a day; hence one of its common names, sea cow. Because of its appetite it has long been eyed as a potential agent for waterweed control. Waterweeds create serious problems for tropical agencies responsible for public health, fisheries, irrigation, navigation, and hydropower generation. The cost of weed control in the waterways of Florida alone is more than fifteen million dollars a year. The problem is aggravated when commercial fertilizers drain from sugar plantations and rice fields into the waterways.

At our workshop we discussed the possibility of transplanting wild manatees to canals—where they would serve as weed-control, slave-labor forces. One afternoon we visited Three Friends Canal, a clean, palm-fringed canal that had been kept free of weeds for fifteen years by two manatees planted in its waters, and a third born there. "One manatee," claimed our native guide, "will mow three or four acres a year."

However, a similar transplant made in Florida in 1964–65 had been a failure. Seven manatees netted in the Miami River were dumped into a canal; all were dead within two years from disease or vandalism.

We also discussed the need for an international center for manatee research. It would foster the conservation of manatees in forty or more countries; it would study the value of manatees for weed control; and it would use the animals as unique, large-bodied, tropical "guinea pigs" in basic biological research. The center would experiment with domesticating sea cows, like cattle, for use as human food (though hardly for milk).

33. A manatee catcher, Datakaran Jeetlal, calms a captive before releasing it to control aquatic weeds. (Photograph: Republic of Guyana, Ministry of Information and Culture)

The workshop attendees enjoyed meeting "Mister Manatee"—Datakaran Jeetlal—a native in his sixties who had become an authority on the habits of manatees. At lunch, he told me that he had captured and transplanted more than seventy of these animals—a record that deserves a place in the *Guinness Book of World Records*. He led us to a place called the Garden of Eden, where we saw a live female lying on a grassy bank. Weighing eight hundred to nine hundred pounds, she had been netted moments earlier in a nearby canal. We were thrilled at this opportunity to see and touch a live manatee, for wild ones, habitually feeding as they do in murky waters, are almost impossible to see and impossible to photograph.

On the following day we traveled by boat up the Abary River to look for wild manatees, but saw none. We did encounter a flock of hoatzins, called by our boatman "Stinkin' Annies." These primitive, fowl-like birds are among the few birds that have fingernails. The young hoatzin, for a short period in its early life, has distinct nails or claws on the first and second digits of each wing which help it to climb before it is able to fly. We saw the hoatzins at dusk just as they were settling clumsily for sleep into a tree at the river's edge.

Then, all at once in the falling night, we were traveling in a convoy of bats, big fellows that swooped alongside our boat and left trails on the surface of the river. I could not see them clearly but supposed they must be fisherman bats *(Noctilio)* engaged in raking the water with their long, sharp claws. A similar bat *(Pizonyx)* lives on Tiburón Island, in the Sea of Cortez, where it feeds on tiny marine fishes.

After the workshop had ended I took a motor-launch trip with zoologist Sandra Husar up a tributary creek of the Demarara River. William Hudson's *Green Mansions* came vividly to mind as we turned and twisted through narrow channels curtained with vines. Iridescent blue morpho butterflies such as I had seen only in curio shops flashed in the dark corridors of the jungle. Everywhere damp and decay, although (to a zoologist) a wholesome sort of decay.

. . . A sudden, rhythmic, thwacking sound, its source revealed as we turned a corner and saw an Indian girl pounding her wet laundry with a paddle . . . A profusion of pinky-purple orchids resting on green moss in the crotch of a tree . . . Another Indian, this one a man gliding along in a dugout canoe piled high with sticks, perhaps the framework for a new hut.

In 1978 the Guyanian jungle came greenly and hotly into mind again as I looked at news photos of the Jonestown suicides.

I shall end on this note. Too long have I lived in colder lands to be fully at ease in the tropics, though I sense their fascination and appreciate their value to the zoologists of the world.

10

Supervising

IN the movie *Little Big Man* Dustin Hoffman reviews his past as a series of stages—the fighting stage, the women stage, the drinking stage, and so on. The life of most zoologists, although less intense, is likewise discontinuous. I am now at the point of describing my administrative or supervisory stage. It will end in a three-year tour of duty as chairman of the United States Marine Mammal Commission. But it is best that I start by describing a certain vision that led me to that stage.

Dreaming of a Sea-mammal Center

I began in the 1940s to dream of a national center devoted to sea-mammal research. It would dispel the persistent clouds of ignorance about creatures known widely as circus seals, as sources of red meat for pet-food processors, as enemies of salmon to be shot on sight, or as smelly, flyblown carcasses to be cleared from bathing beaches. The dolphins and porpoises of American waters were entirely without legal protection. People knew little and cared less about the lives of these graceful animals. Television had not yet brought Flipper and the other aquatic friends of Jacques Cousteau and Marlin Perkins into our living rooms.

The national sea-mammal center of my dreams would be located somewhere on the Pacific Coast, for at least 90 percent of American sea mammals live in Pacific and adjacent Arctic waters. Its zoological staff would discover and develop methods of studying the food habits, reproduction, migration, diving-and-swimming physiology, diseases, care in captivity, and other aspects of sea-mammal biology. The new methods would be applied by wildlife managers living in seacoastal states, by curators of aquariums, and by university researchers. The zoologists of the dream center would not compete with zoologists of other agencies in the collecting of routine information, such as census data, but would concentrate on creative research.

The center would also serve as a data bank. Facts about sea mammals would be kept in accessible form so that they could quickly be retrieved and released to professionals and nonprofessionals alike. The center's educational services would, by feedback, help to win and to hold popular support for the center itself.

But my dream was destined to be only that. It faded as soon as I admitted to myself that neither a government nor a private agency was ever likely to fund a permanent center to the tune of several million dollars a year simply for the general good. However, as specific and urgent national needs for information on sea mammals began to increase in the 1950s and 1960s, many separate agencies did accept responsibility for meeting them. My "center" became many centers, which today are supported by federal, state, municipal, and private organizations.

Among the federal organizations concerned with research and education on sea mammals are the Naval Ocean Systems Center, Naval Oceanographic Office, National Marine Fisheries Service, Marine Mammal Commission, Fish and Wildlife Service, National Science Foundation, National Museum of Natural History, and even the Geological Survey, which carries out research on fossil mammals. Among the state organizations are the universities and the fish-and-game departments of the twenty-two coastal states, with Alaska and California in the lead. Among the city and private organizations are the research facilities of the larger aquariums such as Sea World (San Diego), Sea Life Park (Waimanalo), Marineland (St. Augustine), the New York Aquarium (Brooklyn), and the Steinhart Aquarium (San Francisco).

SUPERVISING

The Early Conferences on Sea Mammals

The same growing interest in sea mammals that gave rise to the foregoing organizations sparked a series of national meetings.
The First International Symposium on Cetacean Research was held in Washington in August 1963 under the sponsorship of the American Institute of Biological Sciences. Attended by scientists from nine countries, it brought together fifty men and women, many of whom had known each other only through correspondence or through exchange of cetacean literature. Perhaps its greatest contribution was that it permitted the sharing of information on instrumentation, that is, the perfection of tools for marking, capturing, and holding whales, dolphins, and porpoises and of studying their life processes. The report of the symposium fills more than eight hundred pages.

In 1964 the Conferences on Biological Sonar and Diving Mammals were inaugurated at Stanford Research Institute (SRI) in Menlo Park, California, on the initiative of Thomas C. Poulter. Tom, then senior scientific adviser at SRI, was a man of many talents. He served as second-in-command to Admiral Byrd on the Second Antarctic Expedition, 1933–35, in which he led the dangerous winter-night trip to rescue the admiral. He later became an expert in wave physics and obtained more than seventy patents dealing with explosives, hydraulic pressures, and seismic exploration. In the late 1950s he had visited Año Nuevo Island, off the central California coast, with intent to erect a missile-tracking station there. But as he clambered over the rocks of that bird-and-seal preserve he saw a sea lion, blind but evidently healthy. "I wondered," Tom told me later, "how the animal could find its food. Do sea lions, like dolphins, echo-range on fish by means of subaquatic sound?"

Tom's lively curiosity, coupled with his long experience in sound recording, led him to call the first conference. It was an immediate success and was the first of nine such meetings. Sea-mammal research in the United States was greatly stimulated by the conferences, for they provided a forum in which zoologists, engineers, and students could talk

34. Thomas C. Poulter gets acquainted with an orphan Steller sea lion; he will later study its behavior. Menlo Park, California, 1962. (Photograph: Stanford Research Institute)

about mutual problems, research methods, and new scientific findings.

SRI sponsored the conferences from 1964 to 1970, at which time Tom launched the Biological Sonar Laboratory as an independent, nonprofit organization. This laboratory sponsored the ninth and last conference (in 1972) at Fremont, California, then closed its doors in 1974 and turned its files over to Kenneth S. Norris, zoologist at the University of California at Santa Cruz.

During the Sixth Conference (in 1969), the International Association of Aquatic Animal Medicine ("Eye Triple A M") was founded. It had been conceived in 1967 at an informal meeting of about thirty veterinarians, zoologists, and aquarium curators

SUPERVISING

at Kissimmee, Florida. Now a lively organization, its interests are centered in the welfare of captive aquatic birds, reptiles, and mammals.

In 1975 Ken Norris and Deborah Day organized the First Annual Conference on the Biology and Conservation of Marine Mammals—the spiritual descendant of the Poulter conferences. It lasted four days, during which time delegates from the United States and abroad heard nearly seventy talks dealing with physiology, behavior, natural history, population changes, and conservation and management.

Also in 1975 a one-day Symposium on Advances in Systematics of Marine Mammals was held at Oregon State University. Specialists in evolution, paleobiology, and comparative anatomy speculated on the origins of living and extinct sea mammals. I enjoyed the symposium, for I have always been attracted to those persons who devote their lives to probing the roots of organic diversity. Their efforts resemble the artistic process in that both draw heavily on imagination and hard work.

At that symposium Deborah A. Duffield, of the University of California, told of her studies aimed at unraveling the tangled threads of whale evolution. She was then looking at chromosomes—the hereditary units. "From only seven drops of fresh whale blood," she reported, "I am able to construct a chromosome map distinctive for the species and (within limits) for the individual." As a spinoff reward, she found high percentages of freak chromosomes in two stranded whales—a blue and a finback. "I wonder," she asked, "whether I am seeing here a classical effect of outcrossing from small, depleted populations." If so, zoologists have additional evidence that the blue and finback stocks are dangerously low.

In the autumn of 1976 a spectacular meeting—the Scientific Consultation on Marine Mammals—convened in Bergen, Norway. It was the first worldwide gathering of specialists interested in the conservation of *all* sea mammals. It had an impressive list of sponsors: The Working Party on Marine Mammals of the Advisory Committee on Marine Resources Research of the Food and Agricultural Organization of the United Nations—with support from the United Nations Environmental Program, the governments of Australia, Canada, Norway, and the United States, the World Wildlife Fund, and the International Union for the Conservation of Nature and Natural Resources. About two hundred delegates from twenty-six nations talked about the value of sea mammals to mankind, about the status of sea-mammal stocks, and about techniques for conserving them. (Such was the outpouring of documents from Bergen that I paid more than one hundred dollars' tariff to air-express a set of them to Seattle.)

The Second Conference on the Biology of Marine Mammals was held in San Diego in December 1977 and was chaired by Forrest G. Wood, zoologist of the Naval Ocean Systems Center. Forrest is a veteran in sea-mammal research, having begun his career in the world's first oceanarium—Marineland of Florida. Among his contributions is the valuable 1973 book *Marine Mammals and Man: The Navy's Porpoises and Sea Lions,* which describes the early history of attempts to catch sea mammals alive, hold them in good health, train them, and study their behavior.

(I myself have never understood why large-scale sea-mammal studies need to be funded by a military agency dedicated to perfecting the tools of death and destruction, which is emphatically not to say that I do not admire the fine, dedicated Navy zoologists who are engaged in those studies.)

Interest in the Second Conference was more than double that in the First; the Second was attended by 470 men and women who listened to, or read, 150 papers. Incidentally, the percentage of women among all speakers and authors was the same at both conferences—about 15 percent. Sea mammalogy, I regret to say, is still largely a man's field.

Chairman of the Marine Mammal Commission

During the decade which was celebrated at midpoint by Earth Day, 1970, national concern for the welfare and, indeed, the survival of sea mammals was growing at a phenomenal rate among laymen as well as among zoologists. Tens of thousands of Americans were joining public-interest organizations designed to inform others about, or to lobby and to litigate in the

SUPERVISING

interests of, sea mammals. Among them were the American Cetacean Society (founded 1967), Friends of the Sea Otter (1968), Greenpeace Foundation (1970), General Whale (1971), Project Jonah (1972), and Monitor, Incorporated (1972). One common objective of such groups was new legislation, preferably at the federal level, which would put an end to abuses of sea mammals by man.

World stocks of blue whales, for example, had fallen to only 6 percent of their numbers before commercial whaling; stocks of humpback whales to 7 percent. ("The International Whaling Commission," wrote Arthur Bourne, British conservationist, "is an object lesson in what happens to a resource which is left in the hands of those most likely to benefit from it.") American tuna fishermen were deliberately killing and discarding about 300,000 porpoises a year. Although the victims were said to have died "incidentally," they were not unlike the casualties that would be left on a busy street by motorists rushing through a crowd of pedestrians. In the absence of regulation, the managers of some zoos and aquariums were obtaining live specimens with little regard for the preservation of local stocks, and were holding the specimens without proper care. The Navy was accused of using sea mammals as aquatic "kamikaze," or suicide, troops against our enemies in Southeast Asia. The accusers claimed that seals and dolphins were being trained to swim stealthily into enemy harbors and there to thrust impact bombs against the hulls of ships! That rumor spread widely after the publication, in 1969, of Robert Merle's fanciful novel *The Day of the Dolphin* and the subsequent release of the motion picture based on it. What the Navy *did* admit to was spending about a half-million dollars a year for research on sea mammals, plus a secret sum for "exploratory and advanced development programs." Some Alaskan Eskimos, using modern firearms and motorboats, were killing bowhead whales and walruses by inhumane and wasteful methods. Annually some 200,000 newborn harp seals were being clubbed for their skins on the ice off eastern Canada, while about 40,000 subadult fur seals were being clubbed for the same purpose in Alaska. Although public anger over the killings was based more on sentiment than zoology, the anger was nonetheless real and acute.

ADVENTURES OF A ZOOLOGIST

By the summer of 1971 national outrage against the real and imagined abuses of sea mammals was approaching the flash point. Many felt that sea-mammal management was not only inefficient but was blind to social values. It was not only allowing the animals to decline in numbers but was allowing them to be killed, captured, or used immorally. Largely as a result of television programs sympathetic to seals, dolphins, and whales, and of national advertisements paid for by animal welfare societies, Americans were demanding tough national legislation.

Congress at first overacted; it listened sympathetically to several bills which would have banned all commercial taking of, and most research taking of, sea mammals. At that point, Lee M. Talbot (senior scientist) and Russell E. Train (administrator) of the President's Council on Environmental Quality offered a White House bill. Its language was close to that of the Marine Mammal Protection Act, which was finally signed by the President on October 21, 1972. If one cuts through the legal foliage of the MMP Act, one can see four new trends in environmental thinking:

First, where the welfare of sea mammals had long been the concern of the coastal states, it now became the concern of the federal government. Predictable cries of outrage rose from states'-righters and rugged individualists. *Alaska* Magazine called the Act a "classic case wherein emotional eccentrics of the 'anti-kill' world have enlisted the aid of headline blinded legislators to effect legislation of such a stupid nature that if its anti-killing thesis were to be extended to all forms of life, we'd be without meat on our tables, clothes on our backs, and even without the vegetables and fruits that would instead be devoured by the insect world."

Second, it replaced the time-honored commercial goal of maximum sustainable yield population with a new one called optimum sustainable population. Controversy over the meaning of the new goal continues to this day. (Some argue that a "best" population can only be decided by the animals themselves!) No matter. The Act was, in the words of Sidney Holt, staff scientist in the United Nations Food and Agricultural Organization, "the first national legislation to place maintenance of the health of ecosystems as a primary objective of wildlife management."

SUPERVISING

Third, the Act emphasizes new uses for sea mammals. It says that these animals "have proven themselves to be resources of great international significance, esthetic and recreational as well as economic." Animals once regarded as marine rawstuffs are now seen as equally valuable alive.

And fourth, the Act introduces the concept of humaneness. Although federal laws had been passed to prevent the suffering of domestic and laboratory animals, none, to my knowledge, had ever been passed specifically to prevent cruelty in the taking of wild animals. The humaneness provision of the Act annoys some of its critics who whine that, although Congress is clearly justified in dealing with the conservation of animals it has no business ordering the manner in which they shall be taken. That, they say, is a matter of personal choice or of private morality. The provision is not an empty gesture. In 1974 the Secretary of Commerce banned the importation of Cape fur-seal skins from South Africa on the grounds that they had been taken from animals killed inhumanely.

The importance of the MMP Act is comparable to that of the Migratory Bird Treaty Act of 1918, the law that saved America's waterfowl. Although the apparatus of sea-mammal conservation still needs fine tuning, the Marine Mammal Commission, and the Departments of Commerce, State, and the Interior, are steadily concerned with the tuning.

~ In the MMP Act, the structure and duties of a Marine Mammal Commission had been specified. On May 14, 1973, President Nixon named me chairman of the first Commission. The Commission was to be an independent agency of the Executive responsible for keeping a perpetual inventory of marine mammal stocks, for promoting research, and for reviewing the activities of the federal departments which had historically dealt with sea-mammal conservation.

The role of the Commission as watchdog over other agencies is certainly novel and is possibly unique in our federal structure. I am reminded that Bernard DeVoto, historian and conservationist, after studying the national plans of the Army Corps of Engineers, complained that "there does not seem to be either in the Corps . . . or outside it any board, commission, or committee charged with the duty of criticising these vast plans in terms of the social good of the whole nation, or indeed, in terms of eventual, practical, as distinguished from theoretical, overall results. This lack troubles me . . ." Similar thoughts, I imagine, were troubling Congress when it set up the Marine Mammal Commission.

My appointment to the Commission came as a total surprise. Lee Talbot telephoned me from Washington on May 3, 1973, to ask whether I would accept it. Although I had never met Talbot I had read his papers on African big-game ecology. After my initial confusion, I began to understand what being a commissioner would entail and how Talbot of Africa had been drawn into sea-mammal affairs.

My fellow commissioners were A. Starker Leopold, professor of zoology, University of California at Berkeley (an authority on wildlife management) and John H. Ryther, chairman of the department of biology, Woods Hole Oceanographic Institution (an authority on marine ecology). Ryther resigned at the end of six months and was replaced by Richard A. Cooley, chairman of geography and environmental studies at the University of California at Santa Cruz.

Nearly a year elapsed before Cooley was appointed—a delay which can be understood if one recalls that President Nixon was then fighting for his political life. He was about to announce, in November 1974: "I am not a crook." When Leopold's two-year term ended in 1975 he was replaced by Donald D. Siniff, department of ecology and behavioral biology, University of Minnesota. When my own three-year term ended in 1976 I was replaced by Douglas G. Chapman, dean of the college of fisheries, University of Washington.

The MMP Act took effect on December 21, 1972. Its first two years were ones of desperation and frustration for those of us who had to carry out its mandates and for others who were involved, in one way or another, with sea mammals. Nearly a year elapsed before Congress, on November 27, 1973, gave the Commission its first money.

In the meantime, the commissioners were being overwhelmed with letters—angry letters from sea-lion collectors under contract to deliver live animals and unable to get the

SUPERVISING

necessary permits; pathetic letters from children who hoped we would save the porpoises from the bad fishermen; incredulous letters from zoologists suddenly aware that they could no longer casually salvage a seal skull from the beach; desperate letters from state game administrators who cried, "For God's sake either manage the sea mammals or return their management to us."

The MMP Act required that we appoint a nine-member committee of scientific advisers on marine mammals. We selected scientists who, as required by the Act, were "knowledgeable in marine mammal ecology and marine mammal affairs." Looking back, I wish that we had appointed one or two who were representative of those known loosely as the protectionists—those who had been most influential in making the MMP Act the law of the land. We might have appointed one or two who were not only scientists but were outspoken champions of the symbolic and social values of sea mammals. We might have sweetened the committee by including a few men or women who (in Theodore Roszak's words, in *Where the Wasteland Ends,* 1972) were identified with "the simple compassion of conservationists and nature lovers." I can think of several who would have qualified—appointees who would have insisted that public sentiment, as well as scientific data, must be considered in making management decisions.

We were lucky to find two highly qualified men, John R. Twiss and Robert Eisenbud, to fill the positions of executive director and general counsel on the Commission's first Washington office staff. They proved to be hard-working, intelligent, and tactful, as well as impervious to the political arrows which were being loosed in volleys from those who felt threatened by the new Act.

When funds for the Commission were at last appropriated, Leopold and I (Ryther having resigned) intensified our search for suitable headquarters. However, we moved too slowly for John D. Dingell, one of the congressmen who had written the MMP Act. I faced him at an oversight hearing in Washington on January 17, 1974, at which time he launched into a long tirade that may have impressed his constituents in Michigan but did little to help the struggling Commission. Explaining the role of the Commission he scolded, "This is not an empty honorific position, where you go about making speeches, wearing robes and attending graduation ceremonies. This is a functioning agency in the government."

Having devoted twenty hours a week for many months to the problems of the Commission, thus far without pay, I felt that his remarks were inappropriate. But all the pieces fell into place with the appointment of Twiss as executive director. By mid-February 1973 he had found office space in the center of Washington, had hired a skeleton staff, and had begun to file the official records of the Commission.

II

Writing

IN the early 1950s I had begun a library search for a list of the world's pinnipeds—or seals, sea lions, and walruses. I wondered: How many kinds are there and where does each kind live? I found that no such list had been published since E.-L. Trouessart's *Catalogus Mammalium*, published in Berlin, 1897–1905. So I asked the National Science Foundation for funds that would enable me to take a year's leave of absence from the Fish and Wildlife Service to prepare an up-to-date list. I proposed to entitle it *The Seals of the World*. The Foundation turned me down, but a year later, having embellished the proposal and retitled it *The Zoological Significance of the Pinnipedia*, I again asked for funds and this time the Foundation came through with eight thousand dollars, to be available in May 1956.

Seals, Sea Lions, and Walruses

Thereupon Beth, our three children, and I embarked on the S.S. *Franconia* for England. Arrived at Cambridge, I was greeted by my old friend from Pribilof days, Colin Bertram. He had arranged study space for me at Cambridge University's Zoology Museum. For the next seven months I shuttled between that study and the libraries of the University, the Philosophical Society, and the Scott Polar Research Institute.

The Polar Research Institute, of which Colin was then director, was founded in honor of Robert Falcon Scott. It has a priceless library of Arctic and Antarctic materials, including many of the journals and sketches from Scott's last trip, as well as later acquisitions. I had trouble getting used to the Cambridge University Library, for it was splintered into a complex of smaller libraries. And the catalog cards—hundreds of thousands—were not filed in drawers but were pasted in huge, leather-bound books. On one occasion, when I had tracked down an eighteenth-century volume, I found that the pages dealing with seals had never been cut! I slipped into an alcove and opened my pocket knife, hoping that the porter would not see me.

The people we met in Cambridge were most amiable. Sir James Gray, zoologist, invited the Scheffers to his home for Sunday dinner. Our host was pleased that I had read his paper on the swimming ability of dolphins. (He and several other zoologists at the University were then specializing in studies of animal locomotion.) He had learned experimentally that, if the resistance of a speeding, live dolphin were equal to that of a rigid model towed at the same speed, the animal's muscles would need to be far more powerful than they actually are. Called Gray's Paradox, the problem was finally solved when other zoologists calculated that, if the flow of water past the dolphin is near perfectly *smooth,* the muscles need be no stronger than those of ordinary mammals. The dolphin, by perfecting a skin surface of the right flexibility to provide a nonturbulent flow, had solved Gray's Paradox fifty million years before man's time.

As work progressed on the seal list I decided to add information on the evolution and biology of pinnipeds, as well as estimates of their world populations. I found it necessary to visit the British Museum (Natural History) in London. There I met Francis C. Fraser, keeper of the seal collections, a man who has contributed enormously to knowledge of the world's whales and dolphins and can be compared in this respect with America's own great Remington Kellogg. Fraser kindly opened for me the Museum's cabinets housing important specimens, including seal skulls collected by early-nineteenth-century explorers.

I still treasure a letter that Fraser wrote me after I returned to America. A certain theologian claimed to have anatomical proof that the Great Whale did indeed swallow and regurgitate Jonah. "Theologians," countered Fraser, "should not question

WRITING

the omnipotence of the Lord but acknowledge the miraculous nature of the Jonah incident."

Fraser took me one evening to a dinner of the Zoological Society of London. As we sat down he pointed to the cover of the Society's printed menu illustrated with a sketch of a penguin mother and chick. "Do you see anything wrong?" he asked. Fortunately for my standing as a zoologist, I did. The artist had given the chick the right shape and size but adult plumage!

But back to seals. What had been conceived as a simple geographical checklist of the species and subspecies turned into a revision of the entire Order Pinnipedia. I discovered that the seals could be usefully classified into thirty-one species and sixteen subspecies—a total of forty-seven so-called trivial names. (That arrangement was later to be altered at several points by other zoologists as a consequence of the perennial—and healthy—contest between the lumpers and the splitters in systematic zoology.)

Challenged by the fact that it had not been done before, I estimated the world population of each species, arriving at a round total of fifteen to twenty-five million seals. I was later proved wrong (for example) in supposing the Chilean fur seal to be extinct, for its survival on Islas Juan Fernandez, far off the coast of Chile, was verified in 1965. And I underestimated the Antarctic crabeaters, the most abundant of all seals. They evidently number nearer fifteen million than the five million I had supposed.

One of the spinoff rewards of writing the seal book was learning about Novaya Zemlya, Houtmans Abrolhos, Isla Guadalupe, and other remote and fascinating places where pinnipeds live. I was obliged to delve into world atlases, draw up distribution maps, and correspond with foreign zoologists. (One of my correspondents, by the way, was Patrice Paulian, of the Muséum d'Histoire Naturelle, in Paris. Misled by the first name, I addressed my first letter to "Mme. Paulian." The reply set me straight—"Je suis homme *entier.*") I was obliged to read articles in French, Spanish, and German. One of the advantages of marine zoology over terrestrial zoology is that its province is the world ocean, a realm including more than 99 percent of the globe's animated envelope.

By the spring of 1957 I had nearly completed the seal manuscript, so our family returned to Seattle where I could add the finishing touches. Then, on pure speculation, I mailed the work to Stanford University Press, where it was accepted and published in 1958 as *Seals, Sea Lions, and Walruses: A Review of the Pinnipedia.*

~ Writing the seal book led me deeply into a field that I had only skirted before—that of systematic zoology. It has two subdivisions.

The first is *taxonomy,* or study of family trees. This is difficult because it calls for evidence of ancient beginnings, and such evidence is rare. Following the clues of fossils, comparative structures, and embryonic features in order to reconstruct the ancestors of modern species is an awesome challenge. I cannot claim to be a primary taxonomist, only a synthesizer of the works of others.

The second is *nomenclature*, or the bestowing of Latin names that express the twig-branch-trunk relationships of the family trees. This work is less difficult; it is little more than scientific librarianship. I sometimes think that too many of its devotees haggle over who said what and when, like the Roman elders who decided that, because Remus saw six vultures first, and Romulus saw twelve later, the new city should be called Rome.

I recall a provocative talk given near the end of World War II by E. Raymond Hall, one of America's respected systematic zoologists. He called his talk "Zoological Subspecies of Man at the Peace Table." He proposed that the United Nations seriously consider ways of maintaining the purity of the five "subspecies"—Caucasian, American Indian, Mongolian, Negro, and Australian Black, stating:

> To imagine one subspecies of man living together on equal terms for long with another subspecies is but wishful thinking and leads only to disaster and oblivion for one or the other. . . . It would seem to be far better for all of us . . . to recognize the biological differences between the subspecies of man and by providing some areas in which citizenship for one subspecies alone will obtain, apply the available zoological knowledge so as to pro-

mote harmony instead of discord and so lengthen the present interval of peace.

Now, Ray Hall is a thoughtful and generous man. He would never rate one "subspecies" of *Homo sapiens* above another. When he argued for the preservation of human diversity he argued strictly from his background as a systematic zoologist. He seemed not to understand, however, that the human animal, alone among the animals, is willing and able to meet at many levels of social and physical intercourse others of its own species who are different.

The Year of the Whale

In November 1966 I received a letter which was to lead me into a new world—the pleasurable world of trade-book writing. The letter was from Kenneth Heuer, then science editor of Charles Scribner's Sons, and it asked me to try writing a "lively book about whales." He had foreseen, several years before the appearance of the first save-the-whale societies, that conservation of whales was at the point of becoming a national issue. I accepted his challenge and began to write *The Year of the Whale*. It was published in 1969, a few months after my retirement from civil service in the same year.

To my surprise, the book was successful and was later translated into six foreign languages. Why this reception? There are, I believe, three reasonable answers.

First, it caught the crest of a wave of public interest in endangered species and rode that wave into the Age of Environmental Awareness. It appeared just when young environmentalists who had been searching for symbols of vanishing Nature found them in the whale, wolf, eagle, and falcon.

Second, it dealt with a species of age-long fascination to humankind. The sperm whale is an extraordinary animal—huge and mysterious; an example of the far limits of power, beauty, and grace that life can reach. I had not fully appreciated the public's fascination with the sperm whale until I had written about him.

And third, the book told a story—the tale of a little whale calf during his first twelve months of life. The story form has always been the most popular of literary forms. Although *The Year of the Whale* also carried zoological facts to thousands of readers who would have spurned on principle any book professing to be educational, those facts were submerged; the story was the thing.

The Year of the Whale illustrates the style known as fiction based on fact. It is one of several styles which a writer can choose when he or she sets out to interpret for human understanding the thoughts and motivations, the movements and the life processes, of a nonhuman animal. At its best, the result is Nature without tears; at its worst, bunny rabbits wearing pinafores. Some of the incidents in *The Year* are based on real happenings; others *could* have happened but, to the best of my knowledge, did not.

The Year of the Seal

Encouraged to continue writing, I delved into literature on the Alaskan fur seal, recalled memories of my visits to the Pribilof Islands, and produced a work which was published in 1970 as *The Year of the Seal.* It told of twelve months in the life of a fictitious creature, the Golden Seal. It also touched on the activities of the zoologists who study the seals, and on the lives of the Pribilof natives who are the stewards of the great fur-seal herd.

Some press reviewers of *The Year of the Seal* felt that it lacked the mystical quality which they had enjoyed in *The Year of the Whale.* I had to agree. It seems that I had known *too much* about seals! Although the seal book was more factual and complete, it was less image-provoking. As I had written in the whale book: "Moving through a dim, dark, cool, watery world of its own, the whale is timeless and ancient; part of our common heritage and yet remote, awful, prowling the ocean floor a half-mile down under the guidance of powers and senses we are only beginning to grasp." While the seal sits on a rock in its own filth, scratching its molting fur.

WRITING

~ Americans were beginning to voice ever more clearly those opinions that were soon to prod Congress into drafting the Marine Mammal Protection Act of 1972. The "protectionists" and the "exploiters" were beginning to dig opposing trenches. And I, too, was undergoing a change of heart. On June 21, 1970, in a letter published in the *New York Times,* I had vigorously defended the killing of Alaskan fur seals as a "harvest" which was yielding five million dollars a year and was removing "surplus" seals which would otherwise have eaten about 150 tons of seafood a day. I made the point, among others, that humaneness is indefinable. My letter brought a pat on the back from the Fouke (formerly Fouke Fur) Company—the firm which had been processing the sealskins since 1921. But it brought a sharp retort from Gladwin Hill, environmental writer for the *Times.* "Victor Scheffer," he wrote, in *Natural History* Magazine, "[is] the ex-federal biologist who deals in 'herd reduction' and such bureaucratic twaddle as 'humaneness is indefinable.' "

Three years later, in 1973, I appeared on a national television show—ABC's "Make a Wish"—and argued that killing fur seals may *not* be the best way of using them for the benefit of mankind. I hinted, even, that no animal is here on Earth to be of use to mankind. The Fouke Company, reacting sharply, called upon President Nixon to remove me from the Marine Mammal Commission to which I had just been appointed. "It is respectfully submitted," wrote an officer of the company, "that Dr. Scheffer should not be a member of this Commission with such obviously biased opposition concerning the conservation program of seal harvesting." Mr. Nixon did not remove me.

I was slowly beginning to shake off the influence of a narrow, zoological-technological training, which had dictated that wild animals exist to be killed for subsistence, sport, or vanity. I was slowly learning that people can find ways of living with animals without having to kill them. And I was learning that, for many sensitive, thoughtful persons, the highest use of a wild animal is to let it be.

~ I am unable to understand myself to the point of knowing why, in later life, I seem to be spending more and more time pleading the cause of animal rights. Certainly I recall no turning point in my path, no conversion along the road to Damascus. I should like to think that the change represents a normal maturing process similar, perhaps, to that which leads the old hunter to put away his guns in favor of binoculars and camera. At any rate, the process represents the domination of conscience over one's need to conform.

Dabbling in Art

During a brief stay in Colorado in the early 1950s I had sensed a quality in the sunlight which had set me to thinking about art patterns in nature. By patterns I mean the configurations infinite in variety that one can see out-of-doors. They are not Nature's Art, for there is no such thing; they are art sources. I began to understand that art and science have common roots. They depend on imagination, they use a searching approach to truth, and they use symbols—pictures and numbers—for explaining truth. I began to recognize a particular kind of duality which is often called beauty-with-meaning, a blending in the mind of the purely sensuous (or imaginative) and the intellectual (or rational).

Deliberately to seek greater pleasure in beauty by thinking about its meaning is of course nothing new. Around the beginning of the twentieth century, famed zoologist Ernst Haeckel wrote of *Kunstformen der Natur* (Art Forms in Nature), illustrating his thesis with incredibly fine sketches of sea organisms. D'Arcy Wentworth Thompson's 1917 classic, *On Growth and Form*, emphasized that Nature uses her blueprints over and over to guide (for example) the dichotomous branching of a seaweed, or of a metallic dendrite, or of a lightning discharge. She seems to respect architectural economy. (My enjoyment of Thompson's work is the richer for knowing that he and I are Pribilof "graduates." He was sent to the islands in 1896 by the British government to investigate the causes of a decline in the seal population.)

Quite a few modern photographers have interpreted beauty-with-meaning. I mention one of the greatest, Andreas Feininger, *Life* Magazine photographer for twenty years, who in 1977 gave us *The Mountains of the Mind*. Subtitled *A Fantastic Journey into Reality*, it demonstrates (for example) that "the space between the dorsal and ventral carapaces of a horseshoe crab near their juncture [is] a jungle of banyan trees at night, or stalactites and stalagmites in a cave, mysteriously transluminated from within."

During my stay in Colorado I made several hundred color transparencies on 35-millimeter film to illustrate the theme of design in nature. Colorado is beautiful in all months of the year. It is a continuing source of imagery. There one can enjoy the tonic effect of sunlight streaming down through the clean air of the mountains . . . the blazing yellows of cottonwood leaves against shadows . . . the timeless reds of sandstones against deep blue skies . . . the worn granite remnants of the *first* Rocky Mountains standing like paleolithic statues . . . the circles of weather-varnished rocks where Indians staked their tents before the white man came.

Upon returning to Seattle in 1956 I approached a local producer of educational films with the proposal that we jointly make a sound-motion film based on the Colorado transparencies. He agreed, and the outcome was a thirty-minute film released as *Art Sources in Nature*. It shows the transparencies one by one; its only motion is footage of the author introducing and ending the series. Although it was good fun, that cinematographic venture was unprofitable; it netted me thirty-five dollars.

In 1970 I bought a larger camera, a Swedish model using 60-millimeter film, and with it made a new collection of color transparencies. Some of the best were published in 1971 in a book entitled *The Seeing Eye*. There I suggested that the elements of beauty in nature can conveniently be classified as form, texture, and color. One or another is often the dominant message in a pattern, or the three can appear in various combinations.

I tried once again to express by words and pictures the thought that, if one deliberately searches for meaning within beauty, one's enjoyment of the outdoors can be heightened. In 1977 Pacific Search Press published my *Messages from the Shore,* self-described as "a testimonial to saltwater beaches and coasts." It emphasized the theme of evolution—the neverending process which creates the forms, textures, and colors of seashore organisms.

~ A final thought, if I may, on art. The ancient Greeks had no word for it as we now use it to mean fine art or art for art's own sake. Beauty and good they knew, yes, but not art as an act of exploring sensual frontiers. They had no word for art because they needed none. In the Greek mind a work of art was, for example, a vase well proportioned, easily handled, and having a texture pleasing to the touch. Few Greeks would have understood the kinds of arty "false vases," meant only to be enjoyed from a distance, which are to be seen in modern art galleries or on potters' shelves. Edith Hamilton, honorary citizen of Athens and lifelong student of the Greek way of life, wrote, in *The Greek Way*, 1964: "Greek art [our word] is intellectual art, the art of men who were clear and lucid thinkers, and it is therefore plain art. Artists than whom the world has never seen greater, men endowed with the spirit's best gift, found their natural method of expression in the simplicity and clarity which are the endowment of the unclouded reason."

A Voice for Wildlife

I wrote earlier that my 1974 book, *A Voice for Wildlife*, grew out of lectures given to University of Washington forestry students. While I was preparing the lectures I was collecting items of interest about confrontations between animals and people. Some 2,800 kinds of wild birds and mammals live in North America. Man tries to increase the numbers of certain ones that he hunts for sport, fashion, or utility. He tries to decrease the numbers of others that he looks upon as pests because they annoy him or interfere with his agriculture, or threaten his health. With few of them is he able to coexist without trying to change their numbers or their ways of life. All of which means that several thousand men and women are dealing every day with problems generated at the points where wildlife and humankind intersect. These men and women are zoologists and veterinarians, game wardens and game administrators, members of public-interest groups, teachers, and others in a vast complex which manages to keep a sort of balance between humankind and its wild brethren. The process in which they invest their energies is wildlife management.

In the book I tried to describe wildlife management for nonzoologists. I offered a diagrammatic flow-chart of its typical operations. Input is represented by biological and socioeconomic data which enter an administrative structure of state and federal agencies having legal responsibility for wildlife. (These agencies are, roughly, the game departments.) The structure has two outputs. One is termed "dealing with wildlife," either indirectly by manipulating habitats or directly by arranging for hunting seasons, controlling nuisance animals, raising animals on game farms, and exporting or importing animals. The other is termed "dealing with people," by educating the public in wildlife conservation, by advising legislators and courts of law, by enforcing game laws, by training conservation workers, and by negotiating with landowners.

A Voice was illustrated by Ugo Mochi, considered up to the time of his death in 1977 as the world's finest exponent of "shadows in outline," or silhouettes. With a lithographic blade honed to a razor's edge he would cut outline pictures from a piece of black paper placed on heavy glass, then mount the delicate tracery on a white board. His specialty was animals. The American Museum of Natural History prizes one of his murals which, although ten feet high, was cut from a single piece of paper. When he worked on *A Voice* he was eighty-three years old and ill, yet he rendered pine needles and cactus spines as gracefully as though he were drawing them with a fine-pointed pen.

A Natural History of Marine Mammals

In the early 1960s I had begun to lecture with color slides on the sea mammals of the world. I revised the lecture up to 1976, when its central ideas were incorporated in a book entitled *A Natural History of Marine Mammals*. When I started to write it, no similar book had been published, although sea-mammal literature in the form of scientific and popular articles was growing rapidly. In 1882 Joel Asaph Allen's bibliography on cetaceans and sirenians had listed about 3,000 titles. Today I suppose that a complete list of books and articles on sea mammals would contain 50,000 to 100,000 entries.

A Natural History of Marine Mammals was illustrated with lively ink sketches from the pen of artist Peter Parnall. In planning the book I had felt that it should be richly illustrated for the benefit of readers unfamiliar with sea mammals. These vary widely in shape and size from the young sea otter, which at first glance could be taken for a puppy, to the great blue whale, a superbeast whose glistening body reaches a length of one hundred feet and a weight of two hundred tons. . . . Or did before the whalemen came; now the largest blues are gone.

I enjoyed writing *A Natural History* because it gave me further opportunity to deal with evolution, one of the fundamental themes of zoology. One after another, during sixty million years or longer, pioneer stocks of warm-blooded, milk-giving animals left the continents to exploit the food riches of the sea. Those stocks are represented today by six groups of sea mammals: the sea otter, walking seals, crawling seals, sirenians, toothed cetaceans, and baleen cetaceans. As each novel body structure (such as the flipper or modified hand) appeared it demonstrated beautifully in Nature's laboratory the coevolution of body form and body function.

And writing the book gave me an opportunity to describe some of the novel techniques which zoologists have invented to study sea mammals. For example, Nancy Telfer and her coworkers demonstrated that dolphins can, and do, drink sea water. They placed three kinds of dolphins in a pool of sea water containing a harmless radioactive tracer chemical. At two-hour intervals they collected and analyzed samples of urine, blood, and "tears" from the test animals. They found that the tracer content of the body fluids rose steadily; the animals were indeed ingesting the water of their pool.

Dutch anatomist Everhard J. Slijper, in order to estimate the milk production of a large whale, synthesized the following data: the observed periodicity of nursing of an aquarium dolphin, the composition of whale milk, and the growth rate of a suckling whale. He concluded that a large whale may give 130 gallons of milk a day in 40 feedings of 3¼ gallons each.

Australian zoologist Jeanne M. Bow has discovered how to

tell the age of a sperm whale from its teeth. She slices a tooth lengthwise and etches it with formic acid, thus revealing alternate layers of hard and soft ivory. Each pair of layers represents one year's growth.

John C. Lilly, an American scientist, proved that dolphins "sleep"—or something close to it. He stationed two watchers, one on the right side and one on the left side of a dolphin resting in a narrow tub of water. In twenty-four hours, the animal closed both eyes for less than five minutes and closed each eye separately for three or four hours.

~ I end this chapter on zoological writing by pointing to the present need for trained translators of technical research reports. The need is growing exponentially and is not being met. "Today," wrote the late Jacob Bronowski, natural philosopher, in *A Sense of the Future,* 1977, "the scientist's language shares no imagery with the vernacular, and is as private and imprisoned as the modern poet's or the modern painter's."

Especially in the field of forensic zoology is the communication gap widening. Lawmakers and judges are being called upon increasingly to deal with issues involving endangered species or wild animals that cross international boundaries, and with sticky legal definitions such as those of "humaneness" in the taking of wildlife and of "optimum" sustainable populations. Faced with the task of breaking new legal ground—of establishing precedents—the lawmakers and judges turn to the literature of zoology and wildlife management for background information. That literature ought to be, though seldom is, written in language understandable to nonzoologists.

12

A Moral Ending

DURING most of my life I have studied wild animals and have worn the occupational badge of wildlife management zoologist—wildlife management being that profession which aims to provide humankind with a sort of useful abundance of animals. But privately I am a naturalist, meaning one who is curious about the whole natural world. And someday I should like to earn the badge of humanist, although time is running out.

Credo

It is often asked of old naturalists, "What do you believe?"
I believe that we Americans are living in a time of change marked by increasing respect for nature and naturalness and respect for life. Public attitudes are beginning to change as a result of awareness of our dependence upon, and responsibility for, natural ecosystems, and awareness of the value of living organisms, whether human or not.

In a recent article in the *Wildlife Society Bulletin* I suggested as much—and also suggested that wildlife management, like education and medicine, has been weakened by inbreeding. The result is emphasis on structure at the expense of broad helpfulness. We zoologists, I continued, have learned to ask *what, where,* and *how many,* but not *why.* On the margin of that article's manuscript an anonymous referee had written a barnyard obscenity. His reaction led me to think that I might be on the right track.

A MORAL ENDING

Awareness that wildlife populations and their ecosystems are threatened spread with the dawning of the Age of Ecology in the late 1960s. That age was historically marked by the signing of the Endangered Species acts of 1966, 1969, and 1973 and the National Environmental Policy Act of 1969, as well as by the widespread celebration of the first Earth Day in 1970. We suddenly knew that the world is smaller than we had thought.

We, the people, are now committed to the task of understanding the Earth and of healing its wounds—a commitment which means that we shall, with increasing vigor, defend the Earth against the corporate polluters, the strip-miners, the redwood loggers, the pumpers of coastal oil, the dammers of great rivers, and all the other exploiters whose perspective is limited to years rather than centuries. I would add those who chisel the likenesses of men upon granite mountains which, left unscarred, would have spoken far more eloquently of humankind's ability to honor greatness. We, the people, are turning against those who lack vision and imagination.

C. P. Snow expounds the idea of "two cultures," namely those of the scientist and the humanist. But somewhere between the two, or overlapping the two, there is surely a place for the culture of the naturalist and ecologist. There is room for the ideas of John Burroughs, Joseph Wood Krutch, Olaus Murie, Dave Brower, and Rachel Carson. Yes, Rachel was both scientist and humanist. She wrote her master's thesis on "The Development of the Pronephros during the Embryonic and Early Larval Life of the Catfish *(Ictalurus punctatus),"* then went on to write *Silent Spring,* a book that stirred a good part of the civilized world. And just before she died she resolved to her own satisfaction an issue to which, I think, she had given lifelong thought. "It is not half so important," she wrote, in *The Sense of Wonder* (1965), "to *know* as to *feel."*

Naturalists and ecologists are obliged to span both cultures because, in taking an enlightened look at the life systems and the life-support systems of Earth they are obliged to see where people fit in. Educators Paul Shepard and Daniel McKinley believe that ecology is the one discipline most likely to provide *Homo sapiens* with a blueprint for his future. They wrote, in *The Subversive Science: Essays Toward an Ecology of Man:* "Romanticism and primitive mythology which united men with the natural world in the past no longer teach us the unity of life. Scientific conservation, as a benign resourcism, is too narrowly and economically centered."

Awareness that the taking of life can be erosive to the human spirit—that is, can be destructive even to those who are not among the casualties—spread during the long involvement of Americans in the bombing of Southeast Asians. Sensitive persons draw no moral distinction, except in degree, between killing people in political wars and killing wild animals for lubricating oil, soap, luxury goods, or simply for sport. They equate human body counts with bag limits.

When citizens feel helpless to prevent the killing of fellow human beings they turn to caring for animals. Between 1960 and 1977 membership in the Humane Society of the United States rose from 25,000 to 70,000; membership in Defenders of Wildlife from 3,500 to 35,000; and membership in the National Audubon Society from 32,000 to 264,000. I don't maintain that concern for life is the only reason why people join protective organizations, but that it is one of the most important reasons.

I believe that, in the future, the priorities with respect to killing wild animals will be, in descending order: for subsistence, for essential and benign research, for protection against animal pests; for sport; and for luxury markets such as the fur trade. Although I have become a nonhunter I have not become an antihunter. I believe simply that sport hunting is unimaginative and is too often a mark of machismo or of personal immaturity. When my sport-hunting friends insist that killing is not the point of the game, I am skeptical, for they are unwilling to stop short of killing.

I believe that we will learn to appreciate seals above sealskins. Advertisements try to convince us that a sealskin coat marks a distinguished woman. Right ... but distinguished either by ignorance or by insensitivity. Already I see widespread public reaction against the clubbing of seals except when it is done by Native Americans to meet their traditional economic and cultural needs. To end commercial sealing for the luxury market will of course oblige sealers to change jobs and will oblige legislators to revise certain laws and treaties. Those accommodations

A MORAL ENDING

will be natural molts in the growth of the American body.
I cannot agree with Jacques Cousteau, who remarked in Seattle during Involvement Day, October 29, 1977, "The harp seal question is entirely emotional. We have to be logical. . . . Those who are moved by the plight of the harp seal could also be moved by the plight of the pig, with which we make our bacon." Emotion, I think, is one of the *better* reasons for deciding not to kill an animal. (A man who recently proposed to breed Dalmatian dogs for the fur trade learned this in a hurry; he was stopped by dog lovers.) Early in my dual association with animals and people I began to see that wildlife management has to be based on emotional as well as rational considerations. Having been hired, so to speak, as a rationalist I was reluctant to face that fact of life. I continued for a long while to believe that, if only we zoologists with our charts and data could persuade the public to think like zoologists, wildlife management decisions would come easy.

As I write, the American Fur Industry, Incorporated, is suing in federal court to delete a provision of the Marine Mammal Protection Act that bans the importation of skins taken from suckling baby seals. The AFI claims that the Act is "clearly based upon emotion and inadequate knowledge and is utterly contrary to 'sound policies of resource management.'" But who is to define "sound policies"—the furriers or the rest of us?

Harp sealing is a most curious ritual. On the twelfth of March 1978, when the sealers sailed from St. John's, Newfoundland, they celebrated an ecumenical religious service followed by entertainment which was funded, in part, by the governments of Newfoundland and Canada.

And then there's the trapping of land fur bearers, such as raccoons and muskrats, a practice not only offensive to many but patently inhumane. Whereas seals are customarily clubbed into quick insensibility, nearly all land fur bearers are taken in steel leghold traps. The steel trap is a vise, attached to a chain and stake, designed to crush the victim's hand or foot and to hold it until the trapper returns. I know about the swollen flesh and the jagged bones protruding from the flesh, for I myself trapped hundreds of animals—either for profit or for research —before I allowed myself to see what I was doing. The American fur market is the cause of more than a hundred million hours of wild animal distress (pain and fear) each year.

I believe that we are turning toward what zoologists call low-consumptive uses of wildlife—uses that range from bird-and-beast watching and looking at wildlife movies to purely cherishing the thought that animals are out there somewhere sharing the Earth with the rest of us.

The strongest evidence of this turn in public taste is new legislation at both state and federal levels. Missouri voters in 1976 approved a general tax for wildlife conservation, upsetting long tradition that conservation must be funded by hunters and trappers. Although only one in ten Americans is a hunter or trapper, this minority has long controlled the management of wild animals which belong (if that is the right word) to all of us. At the national level, Congress is now considering the Nongame Fish and Wildlife Conservation Act of 1978. Whether it passes is immaterial; one like it is certain to pass sooner or later.

The word *nongame*, by the way, was coined by game-department strategists. Squirming under public pressure to manage wildlife democratically—not for hunters and trappers only—they responded by proposing nongame programs designed to protect chipmunks, herons, eagles, harbor seals, porpoises, and other species which rarely have been hunted or trapped. But to classify wildlife by species is the wrong way to meet today's demand for a fresh approach to management. What people want is classification by use. They want management to provide for a broad spectrum of public uses, regardless of species, corresponding to public interest in those uses.

Of course, low-consumptive users of wildlife will have to take responsibility for animal welfare commensurate with their new voice in management. They will have to campaign for camera safaris (wildlife tours) that do not depend on harassment of animals, for wildlife movies that are filmed humanely and truthfully, and for zoos and aquariums that resemble neither prisons nor vaudeville theaters.

Moreover, I believe that whenever we are bothered by animal pests, as we certainly shall continue to be, we must accommodate to the real world. Sea lions and harbor seals, for example, are increasing as a result of new laws against exploiting

A MORAL ENDING

them, with the result that fishermen are complaining of damage to their fishing gear. After one has tried preventive measures such as frightening, screening against, or making the habitat unattractive to a pest, one is left with no choice but to live-trap it or to kill it. And live-trapping, followed by deportation, is only a temporary and seldom a practical solution. Whenever removal of a pest becomes necessary it should be done humanely and should not be coupled with sport. The animal cannot know whether it is being hunted to improve the condition of humankind or to provide sport, or both, but men and women can know the difference.

Zoology and Morality

Looking back on fifty years in zoology, I believe that its practitioners are becoming more humane and more aware of the needs of society. They are learning to weigh the necessity of killing or hurting animals against the morality of doing so. They are moving toward the day when they will give greater weight to moral than to "rational" or "logical" reasons for using animals.

A 1967 expedition to Mexico to study gray whales was newsworthy because one of its party was Paul Dudley White, President Eisenhower's consultant. The party succeeded in firing morphine-drug harpoons into three adult whales. One whale reacted violently, sheared off the shaft of the harpoon, and disappeared. The other two stranded and died within four hours. All the scientists carried excellent credentials, were sponsored by the National Geographic Society, and performed their experiment with the purest of intentions. In talking with two of them later I sensed their genuine distress that the experiment should have resulted in death. Nowadays, an experiment that posed a similar risk to whales would not even be designed, partly because to execute it would contravene laws of the United States and Mexico and partly because the scientists themselves would regard it as socially unacceptable. Morality with respect to uses of whales has changed perceptibly in one decade.

In 1977 the Federation of American Scientists launched a

campaign to improve the lot of wild, captive, and domesticated animals. It acted as though it had decided that our human lives are bound all around and together and forever with the lives of the animals that were present at our creation. Declared a Federation spokesman:

> Within the human species the direction is clear. In this century, the repression of man by man for reasons of race, religion or sex has a defensive and furtive quality that foreshadows its future decline. By comparison, the treatment of other animals by man—the most intellectual animal—is still a closet subject. . . . Scientists ought to be playing a constructive rather than a rearguard role [in helping animals gain their rights]. . . . And where exploitation is required, as in justified experimentation, we [scientists] ought to be able to show that a defense of these practices can coexist with the ongoing application of live compassion.

The Federation's views are shared by the new Institute for the Study of Animal Problems, a division of the Humane Society of the United States. Human beings are ethically responsible for other animals, maintains Michael Fox, the Institute's director, "because they exist" and "because they are valuable"—as so indeed is all life.

Morality evolves out of human insight and *is* human insight. It needs neither scientific nor divine support. (Long ago, talking to a deeply religious lady, I questioned whether standards of right and wrong, virtue and sin, had been "revealed" to humankind. She assured me, "You're just going through a phase.")

Morality is a consensus of ideas about right and wrong. Because it is based on individual feeling it can only be impermanent; it can only be defined with reference to its time. Historically, whenever a majority of civilized persons have agreed that child labor is wrong or that capital punishment is wrong they have voted to abandon these customs. Every moral code evolves through education and persuasion, and right becomes right by popular definition.

Not to say that there aren't a few fundamental ideas at the core of most moralities or norms. These are, among others, that man is the most important species, that we should love our children and respect our aged, that preservation of our Earth is

A MORAL ENDING

the continuing obligation of all, that destructive uses of Earth are wrong, and that the clearest direction of humanity's course is toward the maximum richness of the individual life within the span of years permitted it by its animal clockwork.

A final question is often put to old naturalists: "Are you optimistic?" If the question means, "Do you think the best is behind us or is still ahead?" I reply that I honestly don't know. If it means, "Are you hopeful?" I answer yes—simply because I believe that one ought to be hopeful. Optimism is a form of courage—the noblest of virtues—and can be acquired by practice. So why not practice?

INDEX

Abary River, 157
Adak Island, 29, 97
aerial reconnaissance, 73, 96–97
Africa, 125, 167–68
agar, 80–81
Agattu Island, 21, 31
Akutan Island, 24–25
Alava, Cape, 133
Albatross (ship), 49
Aleutian Islands, 14–33, 92, 104, 144
Aleuts, 28–30, 49–52, 92, 112
Allen, Joel Asaph, 181
Alpha Helix (ship), 144–52
Amchitka Island, 22, 24, 29, 103
American Broadcasting Company, 177
American Cetacean Society, 165
American Fur Industry, 187
American Institute of Biological Sciences, 161
American Museum of Natural History, xii, 181
American Society of Mammologists, xii, 31, 62
Amundsen, Roald, 82, 140
Amundsen-Scott Station, 141
Anchorage, 97
Año Nuevo Island, 161
Antarctica, 6, 91, 138–43, 161, 171–72
aquariums, 121, 144, 160, 164–65, 188
Arctic, 171–72
Arid Lands Ecology Reserve, 125
Armstrong, Port, 25
Aron, William, 67

art, naturalists' interest in, 178–80
Art Sources in Nature (1956), 179
Ashbrook, Frank G., 49, 53
Atka (ship), 142
Atka Village, 30
atomic energy, 29, 87, 140–41
Attu Island and Village, 21, 30, 94
Australia, 21, 26, 163
Automobile Club of Washington, 133–34
Avon sign, 150

Baddeck, 11
Bailey, Vernon, 8
Baird, Spencer Fullerton, 67
Bald Eagle Protection Act, 1940 (16 U.S.C. 668–668d), 10
balloon, 96, 103
Banner, A. Henry, 74–76
Bartholomew, George A., 101
bat, 68, 158
Baw (sea lion), 148
Bay of Waterfalls, 30
bear, brown, 29
Bear (ship), 61
beetles, 116
Bell, Alexander Graham, 11
Bellevue, Washington, 39, 93, 125, 137
Bellingshausen, Fabian G. von, 140
beluga, 122
Bergen, 163
Bergmann principle, 29
Beringia, 122
Bering Island, 68
Bering Sea, 122, 144–52

[195]

INDEX

Bertram, G. C. L. and Kate, 91, 116, 171
biological clocks, 117
Biological Sonar Laboratory, 162
biomes, 8, 125
birds, diversity of, 124–25
birds, types of: albatross, 9–10; auklet, 20, 22, 28; bunting, 60; cormorant, 20; eagle, 9–10, 21, 188; finch, 23, 29, 60; goose, 22; gull, 9, 22, 97; heron, 188; hoatzin, 157–58; kittiwake, 147; longspur, 60; murre, 20, 22, 59–60, 87, 97, 114, 147–48; murrelet, 148; penguin, 142, 173; petrel, 9–10, 21–22, 148; plover, 60; ptarmigan, 7, 17, 22, 28; puffin, 20, 148; raven, 22; sandpiper, 31, 133; screech owl, 14; shearwater, 9, 21; sparrow, 29, 80; teal, 28; tern, 28; tropical species, 124, 153; waterfowl, 15–16, 167; wren, 17
Birnbaum, Z. William, 97
Black Douglas (ship), 75–77, 80, 94–96
blood, 145, 163
Bogoslof Island, 20–21, 147–48
Bolshevism, 135
Bonham, Kelshaw, 74
Bourne, Arthur, 165
Bovee, Charles Wesley, 137
Bow, Jeanne M., 182–83
Bower, Ward T., 48–49, 53–54, 64
Bransfield, Edward, 140
Bretz, J. Harlan, 38
British Graham Land Expedition, 91
British Museum (Natural History), 172
Brockman, C. Frank, 5–6, 13
Bronowski, Jacob, 183
Brooklyn, 75, 160
Brower, David R., 185
Brown, Robert Z., 96
Brown Bear (ship), 15–33
Bryant, Harold C., 6
Buldir Island, 22
Burch, William R., 133
burro, 135
Burroughs, John, 132, 185
Bychkov, V. A., 113–14
Byrd, Richard Evelyn, 6, 140, 161

California Academy of Sciences, xii
Cambridge, 171–73
camera safaris, 188
Cameron, Amos B. and Sarah, 43
Cameron, Mrs. Hugh, 43
Cameron Creek, 42
Canada, 63, 94, 107–8, 110, 121, 163, 165, 187
cannibalism, 44–46
Cannikin bomb test, 29
canoe, 75
caribou, 61
Carlsborg, 43
Carson, Rachel, 185
Cascade Mountains, 35, 125–26
Catalyst (ship), 9–12
Cayman Islands, 126–31; Conservation Association, 131
Chapman, Douglas G., 99, 168
Chapman, Wilbert M., 74–78
Chile, 173
Chilean Memorial, 133
China, 80
Christchurch, 140, 142
chromosomes, 81, 109, 163
Chuginadak Island, 20
Chukchi Sea, 122
civil service, 13
clam, 28
cold, adaptations to (of sea mammals), 121, 141, 144, 152
Cold Bay, 144, 146
colleges: *see* universities (and colleges)
Colorado, 111, 123, 178–79
Columbian White-tailed Deer National Wildlife Refuge, 41
Commander Islands, 64
Conferences on Biological Sonar and Diving Mammals, 161
Conferences on the Biology and Conservation of Marine Mammals, 163–64
contaminants (pollutants), 29, 112, 124
conventions: *see* treaties
Cook, James, 140
Cooley, Richard A., 168
Copalis, 41–42
Corona del Mar, 81
Cortez, Sea of, 158

[196]

INDEX

Couch, Leo K., 16, 34
Cousteau, Jacques, 159, 187
Cox, Allan, 122
cultures," the "two, 185–86
daffodil bulb flies, 4
Dalquest, Walter W., 34–41
Daniels, Paul C., 139
Day, Deborah, 163
deer, 40–41, 44
Deering, Tam and Ivah, 137
Defenders of Wildlife, 186
Demarara River, 158
Denmark, 147–50
Destruction Island, 35
Devil's Graveyard, 73
DeVoto, Bernard, 167
Dingell, John D., 169–70
Discovery Committee, 121
Disney, Walt, 92–93
distribution of life, 124–25
diversity of life, 124
diving ability, 77, 86, 121, 142, 145
dog, 187
dolphins and porpoises, 9, 78, 109, 121–22, 159, 161, 165, 172, 182–83, 188; salt tolerance of, 182; sleep of, 183; swimming speed of, 172
Dorofeev, Sergei V., 97, 111, 113–15
Douglas, William Orville, 133–36
Doyle, A. Conan, 156
Dubos, René, 69
Dufek, George, 140
Duffield, Deborah A., 163
dugong, 91, 116
Dumont d'Urville Station, 141

Earth Day, 69, 164, 185
echo location, 122, 161
Ecological Society of America, 137
Ecologists Union, 137
ecology and ecosystems, 7–8, 28, 61, 120–21, 125, 132, 166, 184–86
Eisenbud, Robert, 169
Elliott, Henry W., 91
Elton, Charles, 125
Elwha River, 43
endangered species, 82–83, 104, 175, 183
Endangered Species Act, 1973 (16 U.S.C. 1531–1543), 104, 185
Endangered Species Conservation Act, 1969 (16 U.S.C. 668), 185
Endangered Species Preservation Act, 1966 (80 Stat. 926), 185
English, Edith Hardin, 137
entomology, 1, 4
Environmental Awareness, Age of, 175
Eskimos, 149–51, 165
Evans, Cape, 142
Evermann, Barton Warren, xii
evolution: and cognition, 70–71; and natural selection, 119; rate of, 35; study of, 182

Farallons, 95
Fassett, Harry Clifford, 49
Federation of American Scientists, 189–90
Feininger, Andreas, 178
Field Museum of Natural History (Chicago), 48
fisheries, 112, 143, 165
Fisheries Research Board of Canada, 108, 121
fishes, distribution of, 82
fishes, types of: Aleutian, 19, 26; Amazon molly, 82; cod, 20; lantern, 107; pupfish, 82; rock greenling, 20; "seal fish," 77; shad, 78; shark, 86, 107; sunfish, 107; tuna, 165
fish and game departments, 160, 188
flea, 39
Flipper (dolphin), 159
flower fields, 60
fluid habitat, adaptations to (of sea mammals), 121
flying squirrel, 68
forest animals, 123
Forks, 43
Fouke Company (until 1968, Fouke Fur Company), 49, 57, 177
fox, blue, 17, 20–23, 29, 60, 112
Fox, Michael, 190
Franconia (ship), 171
Frank, Richard A., 67
Fraser, Francis C., 172–73
Fremont, 162
Friends of the Sea Otter, 165
fruit flies, 81–82

[197]

INDEX

Funter Bay, 77
fur-bearers, land, 34
fur industry, 186–88
fur seal, northern: age, 53, 56, 72, 89–91, 114, 116; "big cows," 85; birth, 65, 101–3; collars on, 89; counting, 53–57; diving ability, 77, 86; drinking, 99–101; food, 64, 73, 77, 107–8; harem, 48, 85, 99–101; health, 88–89, 113; immobilizing, 148; intermingling of stocks, 64, 94–95, 104, 107–8; liver, 86–87; marking (branding and tagging), 53–57, 64, 90, 95–96, 108; mortality, 48, 56, 88; nets as hazard, 104; observing, 99–103; pelage, 115–16; photographing, 93, 97; populations, 48–49, 53, 97–98, 108, 110; reproduction, 48–49, 53, 57–78, 71, 85, 91, 99–103, 108, 110; sealing (killing), 53, 63, 71, 85, 91, 110–12, 114–15, 117, 165–66, 177; taming, 96, 98; teeth, 90–91, 115–16; territoriality, 99–100, 117–19; twins, 101–2; virility, 117–19; weight, 65, 115, 119

Gabrielson, Ira N., 15–16, 48–50
Gambell, 149–51
Garden of Eden (Guyana), 157
Garfield, Billy, 41
General Whale, 165
Gentry, Roger L., 99
Georgetown, 156
Gertrude Island, 79–80
Gibbins, G. Donald, 49
goat, 135–36
gopher, 35–38
Gray, H. Douglas, 22, 33
Gray, Sir James, 172
Grayland, 78
Gray Wolf River, 42
grazing animals, 125, 135
Greeks, 180
Greenpeace Foundation, 165
Green River, 125
Griffin, Donald R., 69
Grinnell, Joseph, 21
growth layers, 116–17
Guadalupe, Isla, 173
Guberlet, John E., 9

Gulf of Alaska, 11, 72, 108
Gun Bay, 129
Gustafson, Carl E., 72
Guthrie, John C., 138–39
Guyana, 153–58
gypsy moth, 2

Haeckel, Ernst, 178
Hall, E. Raymond, 174–75
Hamilton, Edith, 180
Hanford, 125
Hanson, Elmer C., 103
harpoon boats, 107
Harte, Bret, 141
Haswell, W. A., 124
Hatch, Melville H., 17
Hawaii, 136
Hawk II (ship), 71
heat production (metabolism), 152
Hercules (airplane) 140–41
Herrington, William C., 107
Heuer, Kenneth, 175
Hill, Gladwin, 177
Hiroshima, 87
Hobron, Port, 25
Hodikoff, Mike, 30
Hoffman, Dustin, 159
Holt, Sidney, 166
Home Missions Council of North America, 52
Hopkins, David M., 122
Horn, C. B., 41
horse, 125, 135, 145
Houtmans Abrolhos, 173
Hrdlicka, Ales, 30
Hubbs, Carl L., 82–83
Hudson, William, 158
humaneness, 23, 27, 56–57, 77, 167, 183, 186–88, 190
Humane Society of the United States, 186, 190
Humboldt Bay, 71
Humes, Grant, 43
hunting, 136, 186
Hurd, Roy D., 77
Husar, Sandra, 158
Hutchinson Hill, 122

Ickes, Harold L., 47, 73
Imperial Academy of Science (St. Petersburg, Russia), 85

INDEX

Institute for the Study of Animal Problems, 190
instrumentation, importance of, 161
International Association of Aquatic Animal Medicine, 162
International Symposium on Cetacean Research, 161
International Union for the Conservation of Nature and Natural Resources, 164
International Whaling Commission, 67, 99, 165
invertebrates, marine, 19–20, 130
Involvement Day, 187
Irving, Laurence, 144
Iversen, Jan, 147, 151
Ivy, Michael, 138–39

Jackson, H. H. T., 14–15, 24, 34, 66–67
Japan and Japanese, 2, 26–27, 30, 63–64, 66, 73, 75, 80, 89, 107–8, 110, 112, 121
Jeetlal, Datakaran, 157
Johansen, Kjell, 144
Johnson, Ancel M., 117–19
Jonah, 172
Jonestown, 158
Jordan, David Starr, 104
Juan Fernandez, Islas, 173

Kagamil Island, 30, 92
Kalappa, Landes, 42
Kawamura, Akito, 26–27
Kellogg, Remington, xii, 26, 66–68, 172
Kellogg, Winthrop N., 121
Kenyon, Karl W., 94, 96–97, 101, 104
Ketchikan, 11
killing: *see* life, value of
Kincaid, Trevor, 1–3, 12
King (fur seal), 96, 98
Kirkbride, Benny, 96
Kissimmee, 163
Kitovi, 101
Kodaslides, 13
Kodiak, 144
Kooyman, Gerald L., 142
Kraus, Bertram S., 115–16
krill, 143

Krog, John, 148, 151–52
Krutch, Joseph Wood, 185

La Jolla, 80–83
Lake Washington, 12–13
lantern slides, 13, 81–82
La Push, 72–73, 75, 134
Laughlin, William S., 92
Laycock, George, 29
League of Nations, xii
Leith, David, 145–51
lemming, 58
Lenfant, Claude, 145
Leopold, Aldo, 132–33
Leopold, A. Starker, 168–69
Levardsen, Norman O., 84
Lewis and Clark, 18, 40
life, value of, 185–89
life zones, 7–8
Lilly, John C., 183
limnology, 12
Lindsey, Alton A., 6
Little Kiska Island, 30
Lloyd, Ira Phillip, 137
Long Shot bomb test, 29
lungs, 145
Lyons, Eugene T., 88

Macbain, Alastair, 22
McHugh, J. L., 67
McIntyre, Charlie, 41
Mackay, Ian, 4
McKinley, Daniel, 185
McMurdo Station, 140–42
McNeil Island, 79
Macy, Preston P., 73
magnetism, Earth's, 122
Maizuru, 115
Makahs, 71, 75
Maltzeff, Eugene M., 113
mammals, marine: *see* sea mammals
man, "subspecies" of (talk), 174–75
manatee, 91, 121, 153–58
manchineel, 129
mangrove, 129–30
Manzer, James I., 107
Marineland, 121, 160, 164
Marine Mammal Protection Act, 1972 (P.L. 92–522), 166–70, 177, 187
Marsh, George Perkins, 132
Marshall, Robert, 132

[199]

INDEX

Martin, Fredericka, 51
Marunich, 60
Maud (ship), 82
maximum sustainable yield, 84, 111–12, 166
May, Harry W., 57
Menlo Park, 161–62
Merle, Robert, 165
Merriam, C. Hart, 7–8, 14, 18
Merzenich, Michael M., 40
Messages from the Shore (1977), 179
Mexico, 189
Miami River, 156
mice, 14, 68, 126
Migratory Bird Treaty Act, 1918 (40 Stat. 755), 167
milk, sea mammal, 109, 182
Miller, Gerrit S., xii, 67
Miller, Robert C., 9
Milotte, Alfred G. and Elma, 93–94
Mima mounds and Mima Prairie, 35–38, 138
Minard, Clara L., 41
Mineral King Valley, 135
Mirny Station, 141
Miskito Indians, 128
Missouri, 188
Mochi, Ugo, 181
mole, 35
Monitor, Inc., 165
monkey, 5, 69
morality, 184–91
Morgan, Thomas Hunt, 81–82
mountain beaver, 37, 39–40
Muir, John, 132
Mukkaw Bay, 42
mummies, 29–30, 92
Murie, Olaus Johan, 15–33, 43, 61, 117, 134, 185
Murie, Margaret, 31
Murie Islets, 28
Murphy, Robert Cushman, xii
Muséum National d'Histoire Naturelle (Paris, France), 173
Museum of Vertebrate Zoology, 6
musk oxen, 62
muskrats, 44–46

National Audubon Society, 186
National Environmental Policy Act, 1969 (83 Stat. 852), 61
National Geographic Society, 26, 189
National Science Research Council (Guyana), 156
Native Americans, 186
Natural History of Marine Mammals, A (1976), 181–83
Nature Conservancy, 137–38
Nautilus (ship), 96
Neah Bay, 71
Newby, Terrell C., 79–80
Newfoundland, 187
New York Aquarium, 148, 160
New York Zoological Society, 69
New Zealand, 23, 141–42
Nixon, Richard M., 168, 177
Nongame Fish and Wildlife Conservation Act, 1978 (H.R. 10255 of December 7, 1977), 188
Norris, Kenneth S., 162–63
North American Fauna, 14
Northeast Point, 58, 60
North Pacific Fur Seal Commission, 111
North Pacific Fur Seal Conference, 110
North Pole Expedition, 82
Norway, 121, 148, 163
Novaya Zemlya, 173
Nunivak Island, 148, 154

obesity, 101
Odland, George F., 115
Ogliuga Island, 20
Oil City, 134
Okhotsk, Sea of, 113
Olsen, O. Wilford, 88–89
Olympia, 14, 35, 43, 78
Olympic Coast and Peninsula, 44, 71, 73, 133–38
Olympic National Park, 42, 73
Onagawa, 26
optimum sustainable population (of sea mammals), 166, 183
Osborn, Fairfield, 69
Oscar (seal), 144
Osgood, Wilfred Hudson, 48
Ostrov Tuyleniy, 113
otoliths, 109
Oyhut, 41
oysters, 2

INDEX

Pacific Institute of Fisheries and Oceanology (TINRO), 64
Pacific Search Press, 179
Palmer, Nathaniel, 140
Palmer, Theodore S., xii
Panina, Galina K., 113
Parker, George H., 48
Parker, T. J., 124
Parnall, Peter, 182
Paulian, Patrice, 173
Pearl Harbor, 75
Pend Oreille County, 35
Penguin (ship), 49–50
Perkins, Marlin, 159
pests, animal, 22–23, 39, 47, 135–36, 180, 188–89
Peterson, Richard S., 99, 101–3
Pfeiffer, Egbert W., 40
Philosophical Society, 171
photography: aerial, 96–97, 100; "arty," 179–80; motion picture, 92–93, 179; trick of, 130–31
physiological studies, 144–45
pig, 135–36
Pinchot, Gifford, 132
pinnipeds, 171–74; *see also* fur seals; sea lions; seals; walruses
plankton, 9–11, 13, 17–18, 21, 26, 143 (krill)
Pleistocene (Ice Age), 82, 122, 125
Point of Arches, 138
Poland, 143
poles, Earth's, 82, 122, 141
Pompano Beach, 95
population cycles, 45, 58, 61–63
porpoises: *see* dolphins
Poulter, Thomas C., 161–63
Preble, Edward A., 48
Pribilof, Gerassim, 50
Pribilof Islands, 34, 47–67, 71, 73, 77, 84–106, 117–19, 148, 176, 178
Project Jonah, 165
Puget Sound, 79–109
Puyallup, 1, 50

Queen Charlotte Sound, 10
Quilieutes, 71

Rainier, Mount, 5–7, 13, 16, 120, 125
rat, 28

reindeer, 60–63
Remus and Romulus, 174
Rialto Beach, 133
Riedman, Marianne, 109
Robben Island, 113
Robinson, Rex J., 12
Rocky Mountains, 35, 179
Roebling, John A., 75
Roosevelt, Franklin D., 10, 47–48
Roosevelt, Theodore, 132
Roraima, Mount, 153
Ross, James, 140
Ross Sea, 140–42
Roszak, Theodore, 169
Royds, Cape, 142
Ruina, John F., 138–39
Rusk, Dean, 138–39
Russell, W. M. S., 56
Russia (and U.S.S.R.), 63–64, 92, 107, 110–15, 141
Russian Orthodox church, 30
Russo-Japanese War, 63, 113
Ryther, John H., 168–69

Saint Augustine, Florida, 121, 160
Saint George Island, 50, 58, 97
Saint Johns, 187
Saint Matthew Island, 62–63, 149
Saint Paul Island, 50–51, 54, 57, 59, 84, 93, 97, 100, 112, 117, 122
Saint Petersburg, Russia, 68, 85
Sakhalin Island, 63, 113
salt tolerance (of dolphins), 182
Samalga Island, 94
San Diego, 96, 160, 164; Zoo, 96
San Francisco, 95, 160
San Francisco Sea Products Company, 71
San Miguel Island, 94–95
Santa Cruz, 162
Savannah, 75
Scheffer: Brian, 39; Celia Esther, xiii; Mary Elizabeth, 11, 128; Theophilus (author's father), 14, 39, 75; Victor Blanchard (in photographs), 24, 86, 96
Schmitt, Waldo L., 28
Scientific Consultation on Marine Mammals, 163–64
Scott, Robert Falcon, 140, 142, 171
Scott Polar Research Institute, 171

[201]

INDEX

Scott Station, 141
Scribners (Charles Scribner's Sons), 175
Scripps Institution of Oceanography, 80–83
sea cow, Steller, 93, 123
Sea Life Park, 160
sea lions, 20, 58–59, 73, 84–85, 123, 147–48, 161, 188
Seal Island, 63, 113
Seal Island (motion picture), 93
seals, numbers of, 173
Seals, Sea Lions, and Walruses (1958), 171–75
seals, types of, 171–73; Antarctic, 142–43; Arctic, 122–23; Cape fur, 167; Chilean fur, 173; crabeater, 91, 173; elephant, 109; Guadalupe fur, 121; harbor, 78–80, 144–45, 148, 151–52, 188; harp, 187; leopard, 91; monk, 121; northern fur, *see* fur seal, northern; ribbon, 148; spotted (largha), 150; Weddell, 91, 121, 142
sea mammals: abuses of, 165–66; Arctic distribution of, 122–23; center for study of, 159–60; cold, adaptations to, 121, 141, 144, 152; conferences on, 161–64; ecology of, 120–21; Eskimo uses of, 149–51; evolution of, 119, 121–23, 163; fluid habitat, adaptations to, 121; kinds of, xii, 182; literature on, 171, 181–82; milk, 109, 182; optimum sustainable population, 166, 183; research on, 71–73, 120–22, 160, 165; specialists on, xii; three-dimensional space, adaptations to, 121; uses of, 167
sea otter, 23–24, 28, 41–42, 49, 103–6, 121, 123, 144–46, 152, 182
Seattle, 12, 39, 74–76, 94, 103, 108, 111, 115, 137, 144, 174, 179; Fur Exchange, 44; Zoo, 76, 104
"Seattle laboratory," author's, 27, 115–19, 149
Sea World, 160
Seeing Eye, The (1971), 179
Sefton, Nancy, 131
Semisopochnoi Island, 18
Seton, Ernest Thompson, 42

Shephard, Paul, 185
Shine, Ian, 81
Shipley, Donald D., 74
Sholes, William H., 94, 96–97
shrew, 35
Sierra Club, 134
Simeonoff Island, 28
Siniff, Donald D., 168
Sitka Bay, 21
Skagway, 9, 11
Skana (killer whale), 149, 153
Slijper, Everhard J., 182
Slipp, John W., 78
Slwooko, Beda, 150
snail, 28
Snow, C. P., 185
Southern Ocean, 143
Stanford Research Institute, 161–62
Stanford University Press, 174
State Institute for Whale Research, 121
Steinhart Aquarium, 160
Stejneger, Leonhard H., 68, 113
Steller, George Wilhelm, 85, 93–94
Stepetin, Lavrenty, 99, 101
Stone, Christopher D., 135
Sullivan, Walter, 139
Susie (sea otter), 104
Sverdrup, Harald U. and Gudrun, 82
Swanson, Henry, 95
swimming speed (of dolphins), 172
Symposium on Advances in Systematics of Marine Mammals, 163

Taber, Richard D., 124
Tacoma Aquarium, 144
Taholah, 41
Talbott, Lee M., 166, 168
Tanaga Island, 23
Taylor, Walter P., 106
taxonomy, 174; *see also* zoology, systematic
Telfer, Nancy, 182
textbooks, 124
Third Beach, 136
Third World, 143
Thompson, d'Arcy Wentworth, 178
Thoreau, Henry David, 132
three-dimensional space, adaptations to (of sea mammals), 121
Three Friends Canal, 156

INDEX

Tiburón Island, 158
Tokyo, 115
Tønnesen, Rick, 150
Train, Russell E., 166
traps, steel, 187–88
treaties: Antarctic Treaty, 1959 (T.I.A.S. 4780), 138, 140, 143; Convention for the Conservation of Antarctic Seals, 1972 (ratified by U.S.A. September 15, 1976, but not yet in force), 142–43; Convention for the Regulation of Whaling, 1931 (T.S. 880), xii, *and see also* Whaling Treaty Act of 1936; Interim Convention on Conservation of North Pacific Fur Seals, 1957 (T.I.A.S. 3948), 110; Treaty for the Preservation and Protection of Fur Seals, 1911 (T.S. 564), 57, 63–65, 103, 112
Trippensee, Reuben Edwin, 106
Trouessart, E.-L., 171
True, Frederick William, 66–67
Truman, Harry W., 87
Tule Lake, 44–45
tuna fishermen, 165
Twiss, John R., 169–70
"two cultures," the: *see* cultures, the "two"
Tyee, 25

Unalaska, 49, 95
Unimak Pass, 21
United Kingdom, 121
United Nations, 163, 166
United States: Agriculture, Department of, 14, 47; Arms Control and Disarmament Agency, 138; Army Air Force, 89; Army Corps of Engineers, 167; Atomic Energy Commission, 126; Biological Survey, Bureau of, xii, 7, 14–15, 22, 39, 43, 47–48, 50, 66, 106; C.I.A., 139–40; Coast Guard, 24, 26, 62, 73; Commerce, Department of, 47, 167; Congress, 166, 168, 177, 188; Council on Environmental Quality, 166; Economic Ornithology, Section of, 14; Entomology, Bureau of, 4; F.B.I., 115; Fisheries, Bureau of, 24, 47–48, 66; Fish and Wildlife Service, 22, 41, 44, 47–48, 52–53, 62, 66, 73, 78, 80, 94, 104, 106–7, 111, 121, 160, 171; Forest Service, 134; Geological Survey, 122, 160; Indian Affairs, Commissioner of, 52; Indian Claims Commission, 52; Interior, Department of the, 47, 52, 73, 167; Marine Mammal Commission, 99, 150–51, 159–60, 164–70, 177; National Academy of Sciences, 156; National Marine Fisheries Service, 99, 160; National Museum of Natural History (Smithsonian Institution), xii, 8, 14, 18, 24, 28, 40, 43, 66–68, 74, 83, 160; National Park Service, 5–6, 38, 149; National Science Foundation, 62, 138, 160, 171; National Zoological Park (Smithsonian Institution), 68; Navy, Department of the, 29, 76, 97, 160, 164–65; Sport Fisheries and Wildlife, Bureau of, 104; State, Department of, 64–65, 73, 78, 107, 138–39, 167; Supreme Court, 134; Weather Bureau, 96
universities (and colleges): California Institute of Technology, 81; Cambridge, 91, 103, 171–72; College of the Cayman Islands, 120, 126–31; Colorado Agricultural and Mechanical College, 88; Columbia, 81–82; Evergreen State College, 43, 109; Harvard, 48, 91, 145; Johns Hopkins, 99; Massachusetts Institute of Technology, 138; Midwestern, 34; Oregon State, 163; Purdue, 6; Rockefeller, 69; University of Alaska, 52; University of California: at Berkeley, 34, 101, 168, at San Diego, 104, at Santa Cruz, 109, 162, 168; University College, 56; University of Kansas, 34; University of Michigan, 82; University of Montana, 46; University of Puget Sound, 79; University of Southern California, 135; University of Washington, 1, 6, 9, 12, 27, 34, 74–75, 78, 99, 115, 120, 123–26, 137, 144–45, 168, 180; Washington State, 72; Western

[203]

INDEX

Washington State College, 137; Yale, 133

Vancouver Public Aquarium, 149, 153
van Dyke, Henry, 6
Venable, Larry, 133
vertebrates, natural history of, 124–26
Victoria Land, 140
vitamin A, 86–87
Vladivostok, 64, 115
Voice for Wildlife, A (1974), 123, 180–81
Vostochni, 52, 57
Vostok Station, 141

Waimanalo, 160
Walker, Ernest P., 68–71
walrus, 144, 149–54, 165
Walrus Island, 59
Washington: State Game Department, 34; State Museum, 43, 125
waterweeds, control of, 156
weasel, 42
Whale Bay, 25
whales: age of, 27, 183; milk of, 182; research on, 67, 78, 109, 161
whales, types of: beaked, 83, 95, 122; blue, xii, 24, 163, 182; bowhead, 2, 165; Bryde's, 26; false killer, 78; finback, 24, 163; gray, 189; humpback, 24; killer, 149, 153; sperm, 24–26, 119, 121, 175, 183
Whales Research Institute, 26, 121
whaling, xii, 24–27, 67, 165
Whaling Treaty Act of 1936 (P.L. 74–535), 26
White, Paul Dudley, 189
White, Robert M., 67
White Sea, 97

Wilderness Society, 31
Wildflower Acres, 137
wildland preservation, 132–38
wildlife: management, 15, 106, 123, 166, 180–81, 184; nongame, 188
Wildlife Management Institute, 50
Wildlife Society, 31, 62, 106
Wilke, Ford, 74–75, 107, 111, 115
Wilkes Station, 142
Wilkins, Sir Hubert, 82
Witt, Joseph A., 137
wolf, 11, 42–44
Wood, Forrest G., 164
Woods Hole Oceanographic Institution, 168
World Wars: I, 60, 71; II, 28–30, 49–50, 76–78, 103, 113, 174
World Wildlife Fund, 164
writing, author's, 171–83
Wrobel, Sylvia, 81
Wyoming, 134

Yakima Valley, 35
Year of the Seal, The (1971), 176–77
Year of the Whale, The (1969), 71, 175–76

Zapani Reef, 53–56
Zelenogradsk (ship), 113
Zoological Society of London, 173
zoology: author's apprenticeship in, 1–13, 144; baseline data, 28–29; deductive, 122; defined, xi, 1–2; exploratory, 9, 18; forensic, 183; historical interview, technique, 42; nomenclature, 174; social responsibility within, xiii, 134–37, 166, 169, 184–91; systematic, 2, 104, 163, 174–75
zoos, 165, 188